Praise for *When*

"*When* contains a cornucopia of compelling information and insights."
—*The Philadelphia Inquirer*

"Daniel Pink is one of the few non-fiction authors alive today capable of filtering the work of so many scientific minds through his original human stories and onto the page. He is doggedly diligent in his academic research yet his examples are accessible. . . . Like a long walk with a good, funny, wise friend in a leafy park, reading this book is time well spent."
—*Harper's Bazaar*

"Minutes are precious—and easier than ever to waste. . . . College students and business managers alike may find new ways to organize their schedules and ease difficult decisions by using the 'hidden pattern' of time to their advantage."
—*The Wall Street Journal*

"The breadth of the book's scope is impressive. . . . Pink makes a point to end each chapter with takeaway points that readers can apply to their own lives. *When* is engaging, conversational, and tightly edited, making it an easy yet important read."
—Associated Press

"Intriguing stuff written with a light, assured touch."
—*The Guardian*

"Fascinating . . . truly revelatory."
—*Toronto Star*

"Pink should change many people's understanding of timing with this book, which provides insights from little-known scientific studies in an accessible way. . . . By the book's end, readers will be thinking much more carefully about how they divide up their days and organize their routines."
—*Publishers Weekly*

"Helpful tips and insightful solutions."
—*Forbes*

"[*When*] reveals that timing really *is* everything. . . . This marriage of research, stories, and practical application is vintage Pink, helping us use science to improve our everyday lives."
—*BookPage*

"Known for his popular books on motivation and creativity, Pink tackles the science behind how we organize our time and how we should set up the routines of our days."
—*The Washington Post*

"Solid science backed by sensible action points."
—*Kirkus Reviews*

"Illuminating and often surprising."   —*strategy + business*

"A must-read for marketers, especially if you are keen to optimize your career prospects, improve your leadership skills, understand consumer behavior, or plan a successful campaign."   —*American Marketing Association*

"Pink delivers the bad news about our time-based weaknesses with some good news about how to compensate for them. More delightful still, many of these tips involve simply slowing down, taking breaks, and stealing naps. Alas, none of this advice will prevent time from flying by, but at least there are proven ways to fill our hours a bit better."   —*The Wall Street Journal*

"This riveting examination of time and its role in all aspects of our work and personal lives will likely inspire you to restructure your behaviors completely. Broken into three parts and written with Pink's usual humor, the book explores such profound issues as when to quit your job, exercise, nap, marry, graduate, go first, go last, and drink coffee. On every subject, he and his data are powerfully persuasive."   —*Associations Now*

"Helpful, inspiring, and thoughtful advice."   —*Booklist*

"Combining rigorous research with anecdotes and humor, [Pink] examines why most of us are at our best early in the day before fading after lunch and how we can reorganize our lives to perform at our best. . . . His high-energy style is infectious."   —*AudioFile*

WHEN

Also by Daniel H. Pink

Free Agent Nation
A Whole New Mind
The Adventures of Johnny Bunko
Drive
To Sell Is Human
The Power of Regret

DANIEL H. PINK

WHEN

The Scientific Secrets
of Perfect Timing

RIVERHEAD BOOKS · NEW YORK

RIVERHEAD BOOKS
An imprint of Penguin Random House LLC
penguinrandomhouse.com

Photographs © Daniel H. Pink

All charts created by Tanya Maiboroda

The Library of Congress has catalogued the Riverhead hardcover edition as follows:

Names: Pink, Daniel H., author.
Title: When : the scientific secrets of perfect timing / Daniel H. Pink.
Description: New York : Riverhead Books, 2017.
Identifiers: LCCN 2017033061 | ISBN 9780735210622 (hardcover) |
ISBN 9780735210646 (epub)
Subjects: LCSH: Time—Psychological aspects. | Time perception.
Classification: LCC BF468.P57 2017 | DDC 153.7/53—dc23
LC record available at https://lccn.loc.gov/2017033061

First Riverhead hardcover edition: January 2018
First Riverhead trade paperback edition: January 2019
Riverhead trade paperback ISBN: 9780735210639
International edition ISBN: 9780525535041
International mass-market edition ISBN: 9780525542780

Printed in the United States of America
6th Printing

BOOK DESIGN BY AMANDA DEWEY

CONTENTS

PART ONE. THE DAY

"Across continents and time zones, as predictable as the ocean tides, was the same daily oscillation—a peak, a trough, and a rebound."

"A growing body of science makes it clear: Breaks are not a sign of sloth but a sign of strength."

PART TWO. BEGINNINGS, ENDINGS, AND IN BETWEEN

"Most of us have harbored a sense that beginnings are significant. Now the science of timing has shown that they're even more powerful than we suspected. Beginnings stay with us far longer than we know; their effects linger to the end."

PART THREE.
SYNCHING AND THINKING

Time isn't the main thing. It's the only thing.

—MILES DAVIS

WHEN

INTRODUCTION:
CAPTAIN TURNER'S DECISION

Half past noon on Saturday, May 1, 1915, a luxury ocean liner pulled away from Pier 54 on the Manhattan side of the Hudson River and set off for Liverpool, England. Some of the 1,959 passengers and crew aboard the enormous British ship no doubt felt a bit queasy—though less from the tides than from the times.

Great Britain was at war with Germany, World War I having broken out the previous summer. Germany had recently declared the waters adjacent to the British Isles, through which this ship had to pass, a war zone. In the weeks before the scheduled departure, the German embassy in the United States even placed ads in American newspapers warning prospective passengers that those who entered those waters "on ships of Great Britain or her allies do so at their own risk."[1]

Yet only a few passengers canceled their trips. After all, this liner had made more than two hundred transatlantic crossings without incident. It was one of the largest and fastest passenger ships in the world, equipped with a wireless telegraph and well stocked with lifeboats (thanks in part to lessons from the *Titanic*, which had gone down three years earlier). And, perhaps most important, in charge of the ship was Captain William Thomas Turner, one of the most seasoned seamen in the industry—a gruff fifty-eight-year-old with a career full of accolades and "the physique of a bank safe."[2]

The ship traversed the Atlantic Ocean for five uneventful days. But on May 6, as the hulking vessel pushed toward the coast of Ireland, Turner received word that German submarines, or U-boats, were roaming the area. He soon left the captain's deck and stationed himself on the bridge in order to scan the horizon and be ready to make swift decisions.

On Friday morning, May 7, with the liner now just one hundred miles from the coast, a thick fog settled in, so Turner reduced the ship's speed from twenty-one knots to fifteen knots. By noon, though, the fog had lifted, and Turner could spy the shoreline in the distance. The skies were clear. The seas were calm.

However, at 1 p.m., unbeknownst to captain or crew, German U-boat commander Walther Schwieger spotted the ship. And in the next hour, Turner made two inexplicable decisions. First, he increased the ship's speed a bit to eighteen knots but not to its maximum speed of twenty-one knots, even though his visibility was sound, the waters were steady, and he knew submarines might be lurking. During the voyage, he had assured passengers that he would run the ship fast because at its top speed this ocean liner could easily outrace any submarine. Second, at around 1:45 p.m., in order to calculate his position, Turner executed what's called a "four-point bearing," a maneuver that took forty minutes, rather than carry out a simpler bearing maneuver that would have taken only five minutes. And because of the four-point bearing, Turner had to pilot the ship in a straight line rather than steer a zigzag course, which was the best way to dodge U-boats and elude their torpedoes.

At 2:10 p.m., a German torpedo ripped into the starboard side of the ship, tearing open an immense hole. A geyser of seawater erupted, raining shattered equipment and ship parts on the deck. Minutes later, one boiler room flooded, then another. The destruction triggered a second explosion. Turner was knocked overboard. Passengers screamed and dived for lifeboats. Then, just eighteen minutes after being hit, the ship rolled on its side and began to sink.

Seeing the devastation he had wrought, submarine commander Schwieger headed out to sea. He had sunk the *Lusitania*.

Nearly 1,200 people perished in the attack, including 123 of the 141 Americans aboard. The incident escalated World War I, rewrote the rules of naval engagement, and later helped draw the United States into the war. But what exactly took place that May afternoon a century ago remains something of a mystery. Two inquiries in the immediate aftermath of the attack were unsatisfying. British officials halted the first one so as not to reveal military secrets. The second, led by John Charles Bigham, a British jurist known as Lord Mersey, who had also investigated the *Titanic* disaster, exonerated Captain Turner and the shipping company of any wrongdoing. Yet, days after the hearings ended, Lord Mersey resigned from the case and refused payment for his service, saying, "The *Lusitania* case was a damned, dirty business!"[3] During the last century, journalists have pored over news clippings and passenger diaries, and divers have probed the wreckage searching for clues about what really happened. Authors and filmmakers continue to produce books and documentaries that blaze with speculation.

Had Britain intentionally placed the *Lusitania* in harm's way, or even conspired to sink the ship, to drag the United States into the war? Was the ship, which carried some small munitions, actually being used to transport a larger and more powerful cache of arms for the British war effort? Was Britain's top naval official, a forty-year-old named Winston Churchill, somehow involved? Was Captain Turner, who survived the attack, just a pawn of more influential men, "a chump [who] invited disaster," as one surviving passenger called him? Or had he suffered a small stroke that impaired his judgment, as others alleged? Were the inquests and investigations, the full records of which still haven't been released, massive cover-ups?[4]

Nobody knows for sure. More than one hundred years of investigative reporting, historical analysis, and raw speculation haven't yielded a definitive answer. But maybe there's a simpler explanation that no one has considered. Maybe, seen through the fresh lens of twenty-first-century behavioral and biological science, the explanation for one of the most consequential disasters in maritime history is less sinister. Maybe Captain Turner just made some bad decisions. And maybe those decisions were bad because he made them in the afternoon.

This is a book about timing. We all know that timing is everything. Trouble is, we don't know much about timing itself. Our lives present a never-ending stream of "when" decisions—when to change careers, deliver bad news, schedule a class, end a marriage, go for a run, or get serious about a project or a person. But most of these decisions emanate from a steamy bog of intuition and guesswork. Timing, we believe, is an art.

I will show that timing is really a science—an emerging body of multifaceted, multidisciplinary research that offers fresh insights into the human condition and useful guidance on working smarter and living better. Visit any bookstore or library, and you will see a shelf (or twelve) stacked with books about *how* to do various things—from win friends and influence people to speak Tagalog in a month. The output is so massive that these volumes require their own category: *how-to*. Think of this book as a new genre altogether—a *when-to* book.

For the last two years, two intrepid researchers and I have read and analyzed more than seven hundred studies—in the fields of economics and anesthesiology, anthropology and endocrinology, chronobiology and social psychology—to unearth the hidden science of timing. Over the next two hundred pages, I will use that research to examine questions that span the human experience but often remain hidden from our view. Why do beginnings—whether we get off to a fast start or a false start—matter so much? And how can we make a fresh start if we stumble out of the starting blocks? Why does reaching the midpoint—of a project, a game, even a life—sometimes bring us down and other times fire us up? Why do endings energize us to kick harder to reach the finish line yet also inspire us to slow down and seek meaning? How do we synchronize in time with other people—whether we're designing software or singing in a choir? Why do some school schedules impede learning but certain kinds of breaks improve student test scores? Why does thinking about the past cause us to behave one way, but thinking about the future steer us in a different direction? And, ultimately, how can we build organizations, schools, and lives that take into account the invisible power of timing—that recognize, to paraphrase Miles Davis, that timing isn't the main thing, it's the only thing?

This book covers a lot of science. You'll read about plenty of studies, all of them cited in the notes so you can dive deeper (or check my work). But this is also a practical book. At the end of each chapter is what I call a "Time Hacker's Handbook," a collection of tools, exercises, and tips to help put the insights into action.

So where do we begin?

The place to start our inquiry is with time itself. Study the history of time—from the first sundials in ancient Egypt to the early mechanical clocks of sixteenth-century Europe to the advent of time zones in the nineteenth century—and you'll soon realize that much of what we assume are "natural" units of time are really fences our ancestors constructed in order to corral time. Seconds, hours, and weeks are all human inventions. Only by marking them off, wrote historian Daniel Boorstin, "would mankind be liberated from the cyclical monotony of nature."[5]

But one unit of time remains beyond our control, the epitome of Boorstin's cyclical monotony. We inhabit a planet that turns on its axis at a steady speed in a regular pattern, exposing us to regular periods of light and dark. We call each rotation of Earth a day. The day is perhaps the most important way we divide, configure, and evaluate our time. So part one of this book starts our exploration of timing here. What have scientists learned about the rhythm of a day? How can we use that knowledge to improve our performance, enhance our health, and deepen our satisfaction? And why, as Captain Turner showed, should we never make important decisions in the afternoon?

PART ONE. THE DAY

1.

THE HIDDEN PATTERN
OF EVERYDAY LIFE

What men daily do, not knowing what they do!

—WILLIAM SHAKESPEARE,
Much Ado About Nothing

I f you want to measure the world's emotional state, to find a mood ring large enough to encircle the globe, you could do worse than Twitter. Nearly one billion human beings have accounts, and they post roughly 6,000 tweets every second.[1] The sheer volume of these minimessages—what people say and how they say it—has produced an ocean of data that social scientists can swim through to understand human behavior.

A few years ago, two Cornell University sociologists, Michael Macy and Scott Golder, studied more than 500 million tweets that 2.4 million users in eighty-four countries posted over a two-year period. They hoped to use this trove to measure people's emotions—in particular, how "positive affect" (emotions such as enthusiasm, confidence, and alertness) and "negative affect" (emotions such as anger, lethargy, and guilt) varied over time. The researchers didn't read those half a billion tweets one by one, of course. Instead, they fed the posts into a powerful and widely used computerized text-analysis program called LIWC (Linguistic Inquiry and Word Count) that evaluated each word for the emotion it conveyed.

What Macy and Golder found, and published in the eminent journal *Science*, was a remarkably consistent pattern across people's waking hours. Positive affect—language revealing that tweeters felt active, engaged, and hopeful—generally rose in the morning, plummeted in the afternoon, and climbed back up again in the early evening. Whether a tweeter was North American or Asian, Muslim or atheist, black or white or brown, didn't matter. "The temporal affective pattern is similarly shaped across disparate cultures and geographic locations," they write. Nor did it matter whether people were tweeting on a Monday or a Thursday. Each weekday was basically the same. Weekend results differed slightly. Positive affect was generally a bit higher on Saturdays and Sundays—and the morning peak began about two hours later than on weekdays—but the overall shape stayed the same.[2] Whether measured in a large, diverse country like the United States or a smaller, more homogenous country like the United Arab Emirates, the daily pattern remained weirdly similar. It looked like this:

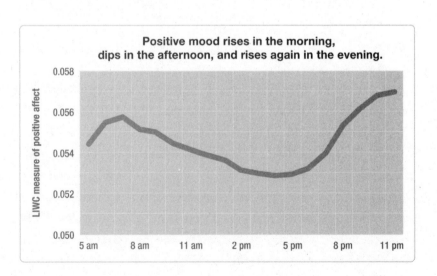

Positive mood rises in the morning, dips in the afternoon, and rises again in the evening.

Across continents and time zones, as predictable as the ocean tides, was the same daily oscillation—a peak, a trough, and a rebound. Beneath the surface of our everyday life is a hidden pattern: crucial, unexpected, and revealing.

Understanding this pattern—where it comes from and what it means—begins with a potted plant, a *Mimosa pudica*, to be exact, that perched on the windowsill of an office in eighteenth-century France. Both the office and the plant belonged to Jean-Jacques d'Ortous de Mairan, a prominent astronomer of his time. Early one summer evening in 1729, de Mairan sat at his desk doing what both eighteenth-century French astronomers and twenty-first-century American writers do when they have serious work to complete: He was staring out the window. As twilight approached, de Mairan noticed that the leaves of the plant sitting on his windowsill had closed up. Earlier in the day, when sunlight streamed through the window, the leaves were spread open. This pattern—leaves unfurled during the sunny morning and furled as darkness loomed—spurred questions. How did the plant sense its surroundings? And what would happen if that pattern of light and dark was disrupted?

So in what would become an act of historically productive procrastination, de Mairan removed the plant from the windowsill, stuck it in a cabinet, and shut the door to seal off light. The following morning, he opened the cabinet to check on the plant and—*mon Dieu!*—the leaves had unfurled despite being in complete darkness. He continued his investigation for a few more weeks, draping black curtains over his windows to prevent even a sliver of light from penetrating the office. The pattern remained. The *Mimosa pudica*'s leaves opened in the morning, closed in the evening. The plant wasn't reacting to external light. It was abiding by its own internal clock.[3]

Since de Mairan's discovery nearly three centuries ago, scientists have established that nearly all living things—from single-cell organisms that lurk in ponds to multicellular organisms that drive minivans—have biological clocks. These internal timekeepers play an essential role in proper functioning. They govern a collection of what are called circadian rhythms (from the Latin *circa* [around] and *diem* [day]) that set the daily backbeat of every creature's life. (Indeed, from de Mairan's potted plant eventually bloomed an entirely new science of biological rhythms known as chronobiology.)

For you and me, the biological Big Ben is the suprachiasmatic nucleus, or SCN, a cluster of some 20,000 cells the size of a grain of rice in the hypothalamus, which sits in the lower center of the brain. The SCN controls the rise and fall of our body temperature, regulates our hormones, and helps us fall asleep at night and awaken in the morning. The SCN's daily timer runs a bit longer than it takes for the Earth to make one full rotation—about twenty-four hours and eleven minutes.[4] So our built-in clock uses social cues (office schedules and bus timetables) and environmental signals (sunrise and sunset) to make small adjustments that bring the internal and external cycles more or less in synch, a process called "entrainment."

The result is that, like the plant on de Mairan's windowsill, human beings metaphorically "open" and "close" at regular times during each day. The patterns aren't identical for every person—just as my blood pressure and pulse aren't exactly the same as yours or even the same as mine were twenty years ago or will be twenty years hence. But the broad contours are strikingly similar. And where they're not, they differ in predictable ways.

Chronobiologists and other researchers began by examining physiological functions such as melatonin production and metabolic response, but the work has now widened to include emotions and behavior. Their research is unlocking some surprising time-based patterns in how we feel and how we perform—which, in turn, yields guidance on how we can configure our own daily lives.

MOOD SWINGS AND STOCK SWINGS

For all their volume, hundreds of millions of tweets cannot provide a perfect window into our daily souls. While other studies using Twitter to measure mood have found much the same patterns that Macy and Golder discovered, both the medium and the methodology have limits.[5] People often use social media to present an ideal face to the world that might mask their true, and perhaps less ideal, emotions. In addition, the industrial-strength analytic tools necessary to interpret so much data can't always detect irony, sarcasm, and other subtle human tricks.

Fortunately, behavioral scientists have other methods to understand what we are thinking and feeling, and one is especially good for charting hour-to-hour changes in how we feel. It's called the Day Reconstruction Method (DRM), the creation of a quintet of researchers that included Daniel Kahneman, winner of the Nobel Prize in Economics, and Alan Krueger, who served as chairman of the White House Council of Economic Advisers under Barack Obama. With the DRM, participants reconstruct the previous day—chronicling everything they did and how they felt while doing it. DRM research, for instance, has shown that during any given day people typically are least happy while commuting and most happy while canoodling.[6]

In 2006, Kahneman, Krueger, and crew enlisted the DRM to measure "a quality of affect that is often overlooked: its rhythmicity over the course of a day." They asked more than nine hundred American women—a mix of races, ages, household incomes, and education levels—to think about the previous "day as a continuous series of scenes or episodes in a film," each one lasting between about fifteen minutes and two hours. The women then described what they were doing during each episode and chose from a list of twelve adjectives (happy, frustrated, enjoying myself, annoyed, and so on) to characterize their emotions during that time.

When the researchers crunched the numbers, they found a "consistent and strong bimodal pattern"—twin peaks—during the day. The women's positive affect climbed in the morning hours until it reached an "optimal emotional point" around midday. Then their good mood quickly plummeted and stayed low throughout the afternoon only to rise again in the early evening.[7]

Here, for example, are charts for three positive emotions—happy, warm, and enjoying myself. (The vertical axis represents the participants' measure of their mood, with higher numbers being more positive and lower numbers less positive. The horizontal axis shows the time of day, from 7 a.m. to 9 p.m.)

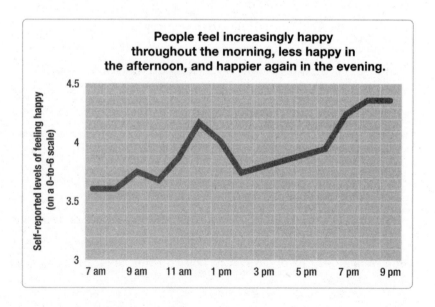

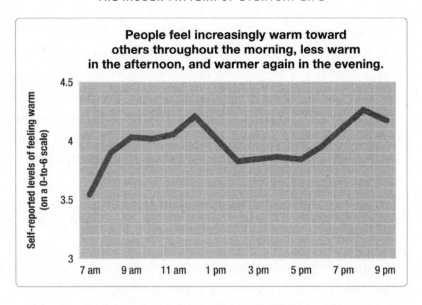

The three charts are obviously not identical, but they all share the same essential shape. What's more, that shape—and the cycle of the day it represents—looks a lot like the one on page 10. An early spike, a big drop, and a subsequent recovery.

On a matter as elusive as human emotion, no study or methodology is definitive. This DRM looked only at women. In addition, *what* and *when* can be difficult to untangle. One reason "enjoying myself" is high at noon and low at 5 p.m. is that we tend to dig socializing (which people do around lunchtime) and detest battling traffic (which people often do in the early evening). Yet the pattern is so regular, and has been replicated so many times, that it's difficult to ignore.

So far I've described only what DRM researchers found about positive affect. The ups and downs of *negative* emotions—feeling frustrated, worried, or hassled—were not as pronounced, but they typically showed a reverse pattern, rising in the afternoon and sinking as the day drew to a close. But when the researchers combined the two emotions, the effect was especially stark. The following graph depicts what you might think of as "net good mood." It takes the hourly ratings for happiness and subtracts the ratings for frustration.

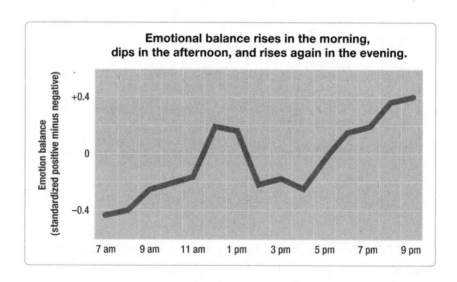

Once again, a peak, a trough, and a rebound.

M oods are an internal state, but they have an external impact. Try as we might to conceal our emotions, they inevitably leak—and that shapes how others respond to our words and actions.

Which leads us inexorably to canned soup.

If you've ever prepared a bowl of cream of tomato soup for lunch, Doug Conant might be the reason why. From 2001 to 2011, Conant was the CEO of Campbell Soup Company, the iconic brand with those iconic cans. During his tenure, Conant helped to revitalize the company and return it to steady growth. Like all CEOs, Conant juggled multiple duties. But one he handled with particular calm and aplomb is the rite of corporate life known as the quarterly earnings call.

Every three months, Conant and two or three lieutenants (usually the company's chief financial officer, controller, and head of investor relations) would walk into a boardroom in Campbell's Camden, New Jersey, headquarters. Each person would take a seat along one of the sides of a long rectangular table. At the center of the table sat a speakerphone, the staging ground for a one-hour conference call. At the other end of the speakerphone were one hundred or so investors, journalists, and, most important, stock analysts, whose job is to assess a company's strengths and weaknesses. In the first half hour, Conant would report on Campbell's revenue, expenses, and earnings the previous quarter. In the second half hour, the executives would answer questions posed by analysts, who would probe for clues about the company's performance.

At Campbell Soup and all public companies, the stakes are high for earnings calls. How analysts react—did the CEO's comments leave them bullish or bearish about the company's prospects?—can send a stock soaring or sinking. "You have to thread the needle," Conant told me. "You have to be responsible and unbiased, and report the facts. But you also have a chance to champion the company and set the record straight." Conant says his goal was always to "take uncertainty out of an uncertain marketplace. For me, these calls introduced a sense of rhythmic certainty into my relationships with investors."

CEOs are human beings, of course, and therefore presumably subject to the same daily changes in mood as the rest of us. But CEOs are also a stalwart lot. They're tough-minded and strategic. They know that millions of dollars ride on every syllable they utter in these calls, so they arrive at these encounters poised and prepared. Surely it couldn't make any difference—to the CEO's performance or the company's fortunes—*when* these calls occur?

Three American business school professors decided to find out. In a first-of-its-kind study, they analyzed more than 26,000 earnings calls from more than 2,100 public companies over six and a half years using linguistic algorithms similar to the ones employed in the Twitter study. They examined whether the time of day influenced the emotional tenor of these critical conversations—and, as a consequence, perhaps even the price of the company's stock.

Calls held first thing in the morning turned out to be reasonably upbeat and positive. But as the day progressed, the "tone grew more negative and less resolute." Around lunchtime, mood rebounded slightly, probably because call participants recharged their mental and emotional batteries, the professors conjectured. But in the afternoon, negativity deepened again, with mood recovering only after the market's closing bell. Moreover, this pattern held "even after controlling for factors such as industry norms, financial distress, growth opportunities, and the news that companies were reporting."[8] In other words, even when the researchers factored in economic news (a slowdown in China that hindered a company's exports) or firm fundamentals (a company that reported abysmal quarterly earnings), afternoon calls "were more negative, irritable, and combative" than morning calls.[9]

Perhaps more important, especially for investors, the time of the call and the subsequent mood it engendered influenced companies' stock prices. Shares declined in response to negative tone—again, even after adjusting for actual good news or bad news—"leading to temporary stock mispricing for firms hosting earnings calls later in the day."

While the share prices eventually righted themselves, these results are remarkable. As the researchers note, "call participants represent the near embodiment of the idealized *homo economicus*." Both the analysts and the executives know the stakes. It's not merely the people on the call who are listening. It's the entire market. The wrong word, a clumsy answer, or an unconvincing response can send a stock's price spiraling downward, imperiling the company's prospects and the executives' paychecks. These hardheaded businesspeople have every incentive to act rationally, and I'm sure they believe they do. But economic rationality is no match for a biological clock forged during a few million years of evolution. Even "sophisticated economic agents acting in real and highly incentivized settings are influenced by diurnal rhythms in the performance of their professional duties."[10]

These findings have wide implications, say the researchers. The results "are indicative of a much more pervasive phenomenon of diurnal rhythms influencing corporate communications, decision-making and performance across all employee ranks and business enterprises throughout the economy." So stark were the results that the authors do something rare in academic papers: They offer specific, practical advice.

"[A]n important takeaway from our study for corporate executives is that communications with investors, and probably other critical managerial decisions and negotiations, should be conducted earlier in the day."[11]

Should the rest of us heed this counsel? (Campbell, as it happens, typically held its earnings calls in the morning.) Our moods cycle in a regular pattern—and, almost invisibly, that affects how corporate executives do their job. So should those of us who haven't ascended to the C-suite also frontload our days and tackle our important work in the morning?

The answer is yes. And no.

VIGILANCE, INHIBITION, AND THE DAILY SECRET TO HIGH PERFORMANCE

Meet Linda. She's thirty-one years old, single, outspoken, and very bright. In college, Linda majored in philosophy. As a student, she was deeply concerned with issues of discrimination and social justice, and participated in antinuclear demonstrations.

Before I tell you more about Linda, let me ask you a question about her. Which is more likely?

a. Linda is a bank teller.
b. Linda is a bank teller and is active in the feminist movement.

Faced with this question, most people answer (b). It makes intuitive sense, right? A justice-seeking, antinuke philosophy major? That sure sounds like someone who would be an active feminist. But (a) is—and must be—the correct response. The answer isn't a matter of fact. Linda isn't real. Nor is it a matter of opinion. It's entirely a matter of logic. Bank tellers who are also feminists—just like bank tellers who yodel or despise cilantro—are *a subset* of all bank tellers, and subsets can never be larger than the full set they're a part of.* In 1983

---

* We can also explain this with some simple math. Suppose there's a 2 percent chance (.02) that Linda is a bank teller. If there's even a whopping 99 percent chance (.99) that she's a feminist, the probability of her being both a bank teller and a feminist is .0198 (.02 x .99)—which is less than 2 percent.

Daniel Kahneman, he of Nobel Prize and DRM fame, and his late collaborator, Amos Tversky, introduced the Linda problem to illustrate what's called the "conjunction fallacy," one of the many ways our reasoning goes awry.[12]

When researchers have posed the Linda problem at different times of day—for instance, at 9 a.m. and 8 p.m. in one well-known experiment—timing often predicted whether participants arrived at the correct answer or slipped on a cognitive banana peel. People were much more likely to get it right earlier in the day than later. There was one intriguing and important exception to the findings (which I'll discuss soon). But as with executives on earnings calls, performance was generally strong in the beginning of the day, then worsened as the hours ticked by.[13]

The same pattern held for stereotypes. Researchers asked other participants to assess the guilt of a fictitious criminal defendant. All the "jurors" read the same set of facts. But for half of them, the defendant's name was Robert Garner, and for the other half, it was Roberto Garcia. When people made their decisions in the morning, there was no difference in guilty verdicts between the two defendants. However, when they rendered their verdicts later in the day, they were much more likely to believe that Garcia was guilty and Garner was innocent. For this group of participants, mental keenness, as shown by rationally evaluating evidence, was greater early in the day. And mental squishiness, as evidenced by resorting to stereotypes, increased as the day wore on.[14]

Scientists began measuring the effect of time of day on brainpower more than a century ago, when pioneering German psychologist Hermann Ebbinghaus conducted experiments showing that people learned and remembered strings of nonsense syllables more effectively in the morning than at night. Since then, researchers have continued that investigation for a range of mental pursuits—and they've drawn three key conclusions.

First, our cognitive abilities do not remain static over the course of a day. During the sixteen or so hours we're awake, they change—often in a regular, foreseeable manner. We are smarter, faster, dimmer, slower, more creative, and less creative in some parts of the day than others.

Second, these daily fluctuations are more extreme than we realize. "[T]he performance change between the daily high point and the daily low point can be equivalent to the effect on performance of drinking the legal limit of alcohol," according to Russell Foster, a neuroscientist and chronobiologist at the University of Oxford.[15] Other research has shown that time-of-day effects can explain 20 percent of the variance in human performance on cognitive undertakings.[16]

Third, how we do depends on what we're doing. "Perhaps the main conclusion to be drawn from studies on the effects of time of day on performance," says British psychologist Simon Folkard, "is that the best time to perform a particular task depends on the nature of that task."

The Linda problem is an analytic task. It's tricky, to be sure. But it doesn't require any special creativity or acumen. It has a single correct answer—and you can reach it via logic. Ample evidence has shown that adults perform best on this sort of thinking during the mornings. When we wake up, our body temperature slowly rises. That rising temperature gradually boosts our energy level and alertness—and that, in turn, enhances our executive functioning, our ability to concentrate, and our powers of deduction. For most of us, those sharp-minded analytic capacities peak in the late morning or around noon.[17]

One reason is that early in the day our minds are more vigilant. In the Linda problem, the politically tinged material about Linda's college experiences is a distraction. It has no relevance in resolving the question itself. When our minds are in vigilant mode, as they tend to be in the mornings, we can keep such distractions outside our cerebral gates.

But vigilance has its limits. After standing watch hour after hour without a break, our mental guards grow tired. They sneak out back for a smoke or a pee break. And when they're gone, interlopers—sloppy logic, dangerous stereotypes, irrelevant information—slip by. Alertness and energy levels, which climb in the morning and reach their apex around noon, tend to plummet during the afternoons.[18] And with that drop comes a corresponding fall in our ability to remain focused and constrain our inhibitions. Our powers of analysis, like leaves on certain plants, close up.

The effects can be significant but are often beneath our comprehension. For instance, students in Denmark, like students everywhere, endure a battery of yearly standardized tests to measure what they're learning and how schools are performing. Danish children take these tests on computers. But because every school has fewer personal computers than students, pupils can't all take the test at the same time. Consequently, the timing of the test depends on the vagaries of class schedules and the availability of desktop machines. Some students take these tests in the morning, others later in the day.

When Harvard's Francesca Gino and two Danish researchers looked at four years of test results for two million Danish schoolchildren and matched the scores to the time of day the students took the test, they found an interesting, if disturbing, correlation. Students scored higher in the mornings than in the afternoons. Indeed, for every hour later in the day the tests were administered, scores fell a little more. The effects of later-in-the-day testing were similar to having parents with slightly lower incomes or less education—or missing two weeks of a school year.[19] Timing wasn't everything. But it was a big thing.

The same appears to be true in the United States. Nolan Pope, an economist at the University of Chicago, looked at standardized test scores and classroom grades for nearly two million students in Los Angeles. Regardless of what time school actually started, "having math in the first two periods of the school day instead of the last two periods increases the math GPA of students" as well as their scores on California's statewide tests. While Pope says it isn't clear exactly why this is happening, "the results tend to show that students are more productive earlier in the school day, especially in math" and that schools could boost learning "with a simple rearrangement of when tasks are performed."[20]

But before you go rearranging your own work schedules to cram all the important stuff before lunchtime, beware. All brainwork is not the same. To illustrate that, here's another pop quiz.

Ernesto is a dealer in antique coins. One day someone brings him a beautiful bronze coin. The coin has an emperor's head on one side and the date 544 BC stamped on the other. Ernesto examines the coin—but instead of buying it, he calls the police. Why?

This is what social scientists call an "insight problem." Reasoning in a methodical, algorithmic way won't yield a correct answer. With insight problems, people typically begin with that systematic, step-by-step approach. But they eventually hit a wall. Some throw up their hands and quit, convinced they can neither scale the wall nor bust through it. But others, stymied and frustrated, eventually experience what's called a "flash of illuminance"—*aha!*—that helps them see the facts in a fresh light. They recategorize the problem and quickly discover the solution.

(Still baffled by the coin puzzle? The answer will make you slap your head. The date on the coin is 544 BC, or 544 years before Christ. That designation couldn't have been used then because Christ hadn't been born—and, of course, nobody knew that he would be born half a millennium later. The coin is obviously a fraud.)

Two American psychologists, Mareike Wieth and Rose Zacks, presented this and other insight problems to a group of people who said they did their best thinking in the morning. The researchers tested half the group between 8:30 a.m. and 9:30 a.m. and the other half between 4:30 p.m. and 5:30 p.m. These morning thinkers were more likely to figure out the coin problem . . . in the afternoon. "Participants who solved insight problems during their non-optimal time of day . . . were more successful than participants at their optimal time of day," Wieth and Zacks found.[21]

What's going on?

The answer goes back to those sentries guarding our cognitive castle. For most of us, mornings are when those guards are on alert, ready to repel any invaders. Such vigilance—often called "inhibitory control"—helps our brains to solve analytic problems by keeping out distractions.[22] But insight problems are different. They require *less* vigilance and *fewer* inhibitions. That "flash of illuminance" is more likely to occur when the guards are gone. At those looser moments, a few distractions can help us spot connections we might have missed when our filters were tighter. For analytic problems, lack of inhibitory control is a bug. For insight problems, it's a feature.

Some have called this phenomenon the "inspiration paradox"—the idea that "innovation and creativity are greatest when we are not at our best, at least with respect to our circadian rhythms."[23] And just as the studies of school performance in Denmark and Los Angeles suggest that students would fare better taking analytic subjects such as math in the morning, Wieth and Zacks say their work "suggests that students designing their class schedules might perform best in classes such as art and creative writing during their non-optimal compared to optimal time of day."[24]

In short, our moods and performance oscillate during the day. For most of us, mood follows a common pattern: a peak, a trough, and a rebound. And that helps shape a dual pattern of performance. In the mornings, during the peak, most of us excel at Linda problems—analytic work that requires sharpness, vigilance, and focus. Later in the day, during the recovery, most of us do better on coin problems—insight work that requires less inhibition and resolve. (Midday troughs are good for very little, as I'll explain in the next chapter.) We are like mobile versions of de Mairan's plant. Our capacities open and close according to a clock we don't control.

But you might have detected a slight hedge in my conclusion. Notice I said "most of us." There is an exception to the broad pattern, especially in performance, and it's an important one.

Imagine yourself standing alongside three people you know. One of you four is probably a different kind of organism with a different kind of clock.

LARKS, OWLS, AND THIRD BIRDS

In the hours before dawn one day in 1879, Thomas Alva Edison sat in his laboratory in Menlo Park, New Jersey, pondering a problem. He had figured out the basic principles of an electric lightbulb, but he still hadn't found a substance that worked as a low-cost, long-lasting filament. Alone in the lab (his more sensible colleagues were home asleep), he absentmindedly picked up a pinch of a sooty, carbon-based substance known as lampblack that had been left out for another experiment, and he began rolling it between his thumb and forefinger—the nineteenth-century equivalent of squeezing a stress ball or trying to one-hop paper clips into a bowl.

Then Edison had—sorry to do this, folks—a lightbulb moment.

The thin thread of carbon that was emerging from his mindless finger rolling might work as a filament. He tested it. It burned bright and long, solving the problem. And now I'm writing this sentence, and perhaps you're reading it, in a room that might be dark but for the illumination of Edison's invention.

Thomas Edison was a night owl who enabled other night owls. "He was more likely to be found hard at it in his laboratory at midnight than at midday," one early biographer wrote.[25]

Human beings don't all experience a day in precisely the same way. Each of us has a "chronotype"—a personal pattern of circadian rhythms that influences our physiology and psychology. The Edisons among us are late chronotypes. They wake long after sunrise, detest mornings, and don't begin peaking until late afternoon or early evening. Others of us are early chronotypes. They rise easily and feel energized during the day but wear out by evening. Some of us are owls; others of us are larks.

You might have heard the larks and owls terminology before. It offers a convenient shorthand for describing chronotypes, two simple avian categories into which we can group the personalities and proclivities of our featherless species. But the reality of chronotypes, as is often the case with reality, is more nuanced.

The first systematic effort to measure differences in humans' internal clocks came in 1976 when two scientists, one Swedish, the other British, published a nineteen-question chronotype assessment. Several years later, two chronobiologists, American Martha Merrow and German Till Roenneberg, developed what became an even more widely used assessment, the Munich Chronotype Questionnaire (MCTQ), which distinguishes between people's sleep patterns on "work days" (when we usually must be awake by a certain hour) and "free days" (when we can awaken when we choose). People respond to questions and then receive a numerical score. For example, when I took the MCTQ, I landed in the most common category—a "slightly early type."

However, Roenneberg, the world's best-known chronobiologist, has offered an even easier way to determine one's chronotype. In fact, you can do it right now.

Please think about your behavior during "free days"—days when you're not required to awaken at a specific time. Now answer these three questions:

1. What time do you usually go to sleep?
2. What time do you usually wake up?
3. What is the middle of those two times—that is, what is your mid-point of sleep? (For instance, if you typically fall asleep around 11:30 p.m. and wake up at 7:30 a.m., your midpoint is 3:30 a.m.)

Now find your position on the following chart, which I've repurposed from Roenneberg's research.

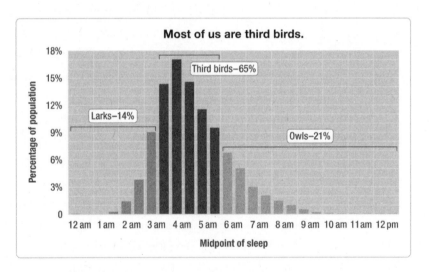

Chances are, you were neither a complete lark nor an utter owl, but somewhere in the middle—what I call a "third bird."* Roenne-

---

* Here's an even simpler method. What time do you wake up on weekends (or free days)? If it's the same as weekdays, you're probably a lark. If it's a little later, you're probably a third bird. If it's much later—ninety minutes or more—you're probably an owl.

berg and others have found that "[s]leep and wake times show a near-Gaussian (normal) distribution in a given population."[26] That is, if you plot people's chronotypes on a graph, the result looks like a bell curve. The one difference, as you can see from the chart, is that extreme owls outnumber extreme larks; owls have, statistically if not physiologically, a longer tail. But most people are neither larks nor owls. According to research over several decades and across different continents, between about 60 percent and 80 percent of us are third birds.[27] "It's like feet," Roenneberg says. "Some people are born with big feet and some with small feet, but most people are somewhere in the middle."[28]

Chronotypes are like feet in another way, too. There's not much we can do about their size or shape. Genetics explains at least half the variability in chronotype, suggesting that larks and owls are born, not made.[29] In fact, the when of one's birth plays a surprisingly powerful role. People born in the fall and winter are more likely to be larks; people born in the spring and summer are more likely to be owls.[30]

After genetics, the most important factor in one's chronotype is age. As parents know and lament, young children are generally larks. They wake up early, buzz around throughout the day, but don't last very long beyond the early evening. Around puberty, those larks begin morphing into owls. They wake up later—at least on free days—gain energy during the late afternoon and evening, and fall asleep well after their parents. By some estimates, teenagers' midpoint of sleep is 6 a.m. or even 7 a.m., not exactly in synch with most high school start times. They reach their peak owliness around age twenty, then slowly return to larkiness over the rest of their lives.[31] The chronotypes of men and women also differ, especially in the first halves of their lives. Men tend toward eveningness, women toward morningness. However, those sex differences begin to disappear around the age of fifty. And as Roenneberg notes, "People over 60 years of age, on average, become even earlier chronotypes than they were as children."[32]

In brief, high school– and college-aged people are disproportionately owls, just as people over sixty and under twelve are disproportionately larks. Men are generally owlier than women. Yet, regardless of age or gender, most people are neither strong larks nor strong owls but are middle-of-the-nest third birds. Still, around 20 to 25 percent of the population are solid evening types—and they display both a personality and a set of behaviors that we must reckon with to understand the hidden pattern of a day.

Let's begin with personality, including what social scientists call the "Big Five" traits—openness, conscientiousness, extraversion, agreeableness, and neuroticism. Much of the research shows morning people to be pleasant, productive folks—"introverted, conscientious, agreeable, persistent, and emotionally stable" women and men who take initiative, suppress ugly impulses, and plan for the future.[33] Morning types also tend to be high in positive affect—that is, many are as happy as larks.[34]

Owls, meanwhile, display some darker tendencies. They're more open and extroverted than larks. But they're also more neurotic— and are often impulsive, sensation-seeking, live-for-the-moment hedonists.[35] They're more likely than larks to use nicotine, alcohol, and caffeine—not to mention marijuana, ecstasy, and cocaine.[36] They're also more prone to addiction, eating disorders, diabetes, depression, and infidelity.[37] No wonder they don't show their faces during the day. And no wonder bosses consider employees who come in early as dedicated and competent and give late starters lower performance ratings.[38] Benjamin Franklin had it right: Early to bed and early to rise makes a person healthy, wealthy, and wise.

Well, not exactly. When scholars have tested Franklin's "gnomic wisdom," they found no "justification for early risers to affect moral

THE HIDDEN PATTERN OF EVERYDAY LIFE

superiority."[39] Those nefarious owls actually tend to display greater creativity, show superior working memory, and post higher scores on intelligence tests such as the GMAT.[40] They even have a better sense of humor.[41]

The problem is that our corporate, government, and education cultures are configured for the 75 or 80 percent of people who are larks or third birds. Owls are like left-handers in a right-handed world—forced to use scissors and writing desks and catcher's mitts designed for others. How they respond is the final piece of the puzzle in divining the rhythms of the day.

SYNCHRONY AND THE THREE-STAGE DAY

Let's return to the Linda problem. Basic logic holds that Linda is less likely to be *both* a bank teller *and* a feminist than she is to be only a bank teller. Most people solve Linda problems more readily at 8 a.m. than at 8 p.m. But some people showed the *reverse* tendency. They were more likely to avoid the conjunction fallacy and produce the correct answer at 8 p.m. than at 8 a.m. Who were these odd-balls? Owls—people with evening chronotypes. It was the same when owls served as jurors in that mock trial. While morning and intermediate types resorted to stereotypes—declaring Garcia guilty and Garner innocent using identical facts—*later* in the day, owls displayed the opposite tendency. They resorted to stereotypes *early* in the day but became more vigilant, fair, and logical as the hours passed.[42]

The ability to solve insight problems, like figuring out that a coin dated 544 BC must be fraudulent, also came with an exception. Larks and third birds had their flashes of illuminance later in the day, during their less optimal recovery stage when their inhibitions

31

had fallen. But Edison-like owls spotted the fraud more readily in the early mornings, *their* less optimal time.[43]

What ultimately matters, then, is that type, task, and time align—what social scientists call "the synchrony effect."[44] For instance, even though it's obviously more dangerous to drive at night, owls actually drive worse early in the day because mornings are out of synch with their natural cycle of vigilance and alertness.[45] Younger people typically have keener memories than older folks. But many of these age-based cognitive differences weaken, and sometimes disappear, when synchrony is taken into account. In fact, some research has shown that for memory tasks older adults use the same regions of the brain as younger adults do when operating in the morning but different (and less effective) regions later in the day.[46]

Synchrony even affects our ethical behavior. In 2014 two scholars identified what they dubbed the "morning morality effect," which showed that people are less likely to lie and cheat on tasks in the morning than they are later in the day. But subsequent research found that one explanation for the effect is simply that most people are morning or intermediate chronotypes. Factor in owliness and the effect is more nuanced. Yes, early risers display the morning morality effect. But night owls are more ethical at night than in the morning. "[T]he fit between a person's chronotype and the time of day offers a more complete predictor of that person's ethicality than does time of day alone," these scholars write.[47]

In short, all of us experience the day in three stages—a peak, a trough, and a rebound. And about three-quarters of us (larks and third birds) experience it in that order. But about one in four people, those whose genes or age make them night owls, experience the day in something closer to the reverse order—recovery, trough, peak.

To probe this idea, I asked my colleague, researcher Cameron French, to analyze the daily rhythms of a collection of artists, writers, and inventors. His source material was a remarkable book, edited by Mason Currey, titled *Daily Rituals: How Artists Work* that chronicles the everyday patterns of work and rest of 161 creators, from Jane Austen to Jackson Pollock to Anthony Trollope to Toni Morrison. French read their daily work schedules and coded each element as either heads-down work, no work at all, or less intense work—something close to the pattern of peak, trough, and recovery.

For instance, composer Pyotr Ilich Tchaikovsky would typically awaken between 7 a.m. and 8 a.m., and then read, drink tea, and take a walk. At 9:30, he went to his piano to compose for a few hours. Then he broke for lunch and another stroll during the afternoon. (He believed walks, sometimes two hours long, were essential for creativity.) At 5 p.m., he settled back in for a few more hours of work before eating supper at 8 p.m. One hundred fifty years later, writer Joyce Carol Oates operates on a similar rhythm. She "generally writes from 8:00 or 8:30 in the morning until about 1:00 p.m. Then she eats lunch and allows herself an afternoon break before resuming work from 4:00 p.m. until dinner around 7:00."[48] Both Tchaikovsky and Oates are peak-trough-rebound kinds of people.

Other creators marched to a different diurnal drummer. Novelist Gustave Flaubert, who lived much of his adult life in his mother's house, would typically not awaken until 10 a.m., after which he'd spend an hour bathing, primping, and puffing his pipe. Around 11, "he would join the family in the dining room for a late-morning meal that served as both his breakfast and lunch." He would then tutor his niece for a while and devote most of the afternoon to resting and reading. At 7 p.m. he would have dinner, and afterward, "he sat and talked with his mother" until she went to bed around 9 p.m. And then he did his writing. Night owl Flaubert's day moved in an opposite direction—from recovery to trough to peak.[49]

After coding these creators' daily schedules and tabulating who did what when, French found what we now realize is a predictable distribution. About 62 percent of the creators followed the peak-trough-recovery pattern, where serious heads-down work happened in the morning followed by not much work at all, and then a shorter burst of less taxing work. About 20 percent of the sample displayed the reverse pattern—recovering in the mornings and getting down to business much later in the day à la Flaubert. And about 18 percent were more idiosyncratic or lacked sufficient data and therefore displayed neither pattern. Separate out that third group and the chronotype ratio holds. For every three peak-trough-rebound patterns, there is one rebound-trough-peak pattern.

So what does this mean for you?

At the end of this chapter is the first of six Time Hacker's Handbooks, which offer tactics, habits, and routines for applying the science of timing to your daily life. But the essence is straightforward. Figure out your type, understand your task, and then select the appropriate time. Is your own hidden daily pattern peak-trough-rebound? Or is it rebound-trough-peak? Then look for synchrony. If you have even modest control over your schedule, try to nudge your most important work, which usually requires vigilance and clear thinking, into the peak and push your second-most important work, or tasks that benefit from disinhibition, into the rebound period. Whatever you do, do not let mundane tasks creep into your peak period.

If you're a boss, understand these two patterns and allow people to protect their peak. For example, Till Roenneberg conducted experiments at a German auto plant and steel factory in which he rearranged work schedules to match people's chronotypes to their work schedules. The results: greater productivity, reduced stress, and higher job satisfaction.[50] If you're an educator, know that all times are not created equal: Think hard about which classes and types of work you schedule in the morning and which you schedule later in the day.

Equally important, no matter whether you spend your days making cars or teaching children, beware of that middle period. The trough, as we're about to learn, is more dangerous than most of us realize.

Time Hacker's Handbook

· CHAPTER 1 ·

HOW TO FIGURE OUT YOUR DAILY WHEN:
A THREE-STEP METHOD

This chapter has explored the science behind our daily patterns. Now here's a simple three-step technique—call it the type-task-time method—for deploying that science to guide your daily timing decisions.

First, determine your chronotype, using the three-question method on page 28 or by completing the MCTQ questionnaire on-line (http://www.danpink.com/MCTQ).

Second, determine what you need to do. Does it involve heads-down analysis or head-in-the-sky insight? (Of course, not all tasks divide cleanly along the analysis-insight axis, so just make the call.) Are you trying to make an impression on others in a job interview, knowing that most of your interviewers are likely to be in a better mood in the morning? Or are you trying to make a decision (whether you should take the job you've just been offered), in which case your own chronotype should govern?

Third, look at this chart to figure out the optimal time of day:

Your Daily When Chart

	Lark	Third Bird	Owl
Analytic tasks	Early morning	Early to midmorning	Late afternoon and evening
Insight tasks	Late afternoon/ early evening	Late afternoon/ early evening	Morning
Making an impression	Morning	Morning	Morning (sorry, owls)
Making a decision	Early morning	Early to midmorning	Late afternoon and evening

For example, if you're a larkish lawyer drafting a brief, do your research and writing fairly early in the morning. If you're an owlish software engineer, shift your less essential tasks to the morning and begin your most important ones in the late afternoon and into the evening. If you're assembling a brainstorming group, go for the late afternoon since most of your team members are likely to be third birds. Once you know your type and task, it's easier to figure out the time.

HOW TO FIGURE OUT YOUR DAILY WHEN: THE ADVANCED VERSION

For a more granular sense of your daily when, track your behavior systematically for a week. Set your phone alarm to beep every ninety minutes. Each time you hear the alarm, answer these three questions:

1. What are you doing?
2. On a scale of 1 to 10, how mentally alert do you feel right now?
3. On a scale of 1 to 10, how physically energetic do you feel right now?

Do this for a week, then tabulate your results. You might see some personal deviations from the broad pattern. For example, your trough might arrive earlier in the afternoon than some people or your recovery may kick in later.

To track your responses, you can scan and duplicate these pages, download a PDF version from my website (http://www.danpink.com /chapter1supplement).

7 a.m.
What I'm doing:
Mental alertness: 1 2 3 4 5 6 7 8 9 10 NA
Physical energy: 1 2 3 4 5 6 7 8 9 10 NA

8:30 a.m.
What I'm doing:
Mental alertness: 1 2 3 4 5 6 7 8 9 10 NA
Physical energy: 1 2 3 4 5 6 7 8 9 10 NA

10 a.m.
What I'm doing:
Mental alertness: 1 2 3 4 5 6 7 8 9 10 NA
Physical energy: 1 2 3 4 5 6 7 8 9 10 NA

11:30 a.m.
What I'm doing:
Mental alertness: 1 2 3 4 5 6 7 8 9 10 NA
Physical energy: 1 2 3 4 5 6 7 8 9 10 NA

1 p.m.
What I'm doing:
Mental alertness: 1 2 3 4 5 6 7 8 9 10 NA
Physical energy: 1 2 3 4 5 6 7 8 9 10 NA

2:30 p.m.

What I'm doing:

Mental alertness: 1 2 3 4 5 6 7 8 9 10 NA

Physical energy: 1 2 3 4 5 6 7 8 9 10 NA

4 p.m.

What I'm doing:

Mental alertness: 1 2 3 4 5 6 7 8 9 10 NA

Physical energy: 1 2 3 4 5 6 7 8 9 10 NA

5:30 p.m.

What I'm doing:

Mental alertness: 1 2 3 4 5 6 7 8 9 10 NA

Physical energy: 1 2 3 4 5 6 7 8 9 10 NA

7 p.m.

What I'm doing:

Mental alertness: 1 2 3 4 5 6 7 8 9 10 NA

Physical energy: 1 2 3 4 5 6 7 8 9 10 NA

8:30 p.m.

What I'm doing:

Mental alertness: 1 2 3 4 5 6 7 8 9 10 NA

Physical energy: 1 2 3 4 5 6 7 8 9 10 NA

10 p.m.

What I'm doing:

Mental alertness: 1 2 3 4 5 6 7 8 9 10 NA

Physical energy: 1 2 3 4 5 6 7 8 9 10 NA

11:30 p.m.

What I'm doing:

Mental alertness: 1 2 3 4 5 6 7 8 9 10 NA

Physical energy: 1 2 3 4 5 6 7 8 9 10 NA

WHAT TO DO IF YOU DON'T HAVE CONTROL OVER YOUR DAILY SCHEDULE

The harsh reality of work—whatever we do, whatever our title—is that many of us don't fully control our time. So what can you do when the rhythms of your daily pattern don't coincide with the demands of your own daily schedule? I can't offer a magic remedy, but I can suggest two strategies to minimize the harm.

1. Be aware.

Simply knowing that you're operating at a suboptimal time can be helpful because you can correct for your chronotype in small but powerful ways.

Suppose you're an owl forced to attend an early-morning meeting. Take some preventive measures. The night before, make a list of everything you'll need for the gathering. Before you sit down at the conference table, go for a quick walk outside—ten minutes or so. Or do a small good deed for a colleague—buy him a coffee or help him carry some boxes—which will boost your mood. During the meeting, be extra vigilant. For instance, if someone asks you a question, repeat it before you answer to make sure you've gotten it right.

2. Work the margins.

Even if you can't control the big things, you might still be able to shape the little things. If you're a lark or a third bird and happen

43

to have a free hour in the morning, don't fritter it away on e-mail. Spend those sixty minutes doing your most important work. Try managing up, too. Gently tell your boss when you work best, but put it in terms of what's good for the organization. ("I get the most done on the big project you assigned me during the mornings—so maybe I should attend fewer meetings before noon.") And start small. You've heard of "casual Fridays." Maybe suggest "chronotype Fridays," one Friday a month when everyone can work at their preferred schedule. Or perhaps declare your own chrono-type Friday. Finally, take advantage of those times when you *do* have control over your schedule. On weekends or holidays, craft a sched-ule that maximizes the synchrony effect. For example, if you're a lark and you're writing a novel, get up early, write until 1 p.m., and save your grocery shopping and dry-cleaning pickup for the after-noon.

WHEN TO EXERCISE: THE ULTIMATE GUIDE

I've focused mostly on the emotional and cognitive aspects of our lives. But what about the physical? In particular, what's the best time to exercise? The answer depends in part on your goals. Here's a simple guide, based on exercise research, to help you decide.

Exercise in the morning to:

- **Lose weight:** When we first wake up, having not eaten for at least eight hours, our blood sugar is low. Since we need blood sugar to fuel a run, morning exercise will use the fat stored in our tissues to supply the energy we need. (When we exercise after eating, we use the energy from the food we've just con-sumed.) In many cases, morning exercise may burn 20 percent more fat than later, post-food workouts.[1]

- **Boost mood**: Cardio workouts—swimming, running, even walking the dog—can elevate mood. When we exercise in the morning, we enjoy these effects all day. If you wait to exercise until the evening, you'll end up sleeping through some of the good feelings.
- **Keep to your routine**: Some studies suggest that we're more likely to adhere to our workout routine when we do it in the morning.[2] So if you find yourself struggling to stick with a plan, morning exercise, especially if you enlist a regular partner, can help you form a habit.
- **Build strength**: Our physiology changes throughout the day. One example: the hormone testosterone, whose levels peak in the morning. Testosterone helps build muscle, so if you're doing weight training, schedule your workout for those early-morning hours.

Exercise in the late afternoon or evening to:

- **Avoid injury**: When our muscles are warm, they're more elastic and less prone to injury. That's why they call what we do at the beginning of our workout a "warm-up." Our body temperature is low when we first wake up, rises steadily throughout the day, and peaks in the late afternoon and early evening. That means that in later-in-the-day workouts our muscles are warmer and injuries are less common.[3]
- **Perform your best**: Working out in the afternoons not only means that you're less likely to get injured, it also helps you sprint faster and lift more. Lung function is highest this time of the day, so your circulation system can distribute more oxygen and nutrients.[4] This is also the time of day when strength peaks, reaction time quickens, hand-eye coordination sharpens, and heart rate and blood pressure drop. These factors

make it a great time to put on your best athletic performance. In fact, a disproportionate number of Olympic records, especially in running and swimming, are set in the late afternoon and early evening.[5]

- **Enjoy the workout a bit more:** People typically perceive that they're exerting themselves a little less in the afternoon even if they're doing exactly the same exercise routine as in the morning.[6] This suggests that afternoons may make workouts a little less taxing on the mind and soul.

FOUR TIPS FOR A BETTER MORNING

1. Drink a glass of water when you wake up.

How often during a day do you go eight hours without drinking anything at all? Yet that's what it's like for most of us overnight. Between the water we exhale and the water that evaporates from our skin, not to mention a trip or two to the bathroom, we wake up mildly dehydrated. Throw back a glass of water first thing to rehydrate, control early morning hunger pangs, and help you wake up.

2. Don't drink coffee immediately after you wake up.

The moment we awaken, our bodies begin producing cortisol, a stress hormone that kick-starts our groggy souls. But it turns out that caffeine interferes with the production of cortisol—so starting the day immediately with a cup of coffee barely boosts our wakefulness. Worse, early-morning coffee increases our tolerance for caffeine, which means we must gulp ever more to obtain its benefits. The better approach is to drink that first cup an hour or ninety minutes after waking up, once our cortisol production has peaked and the caffeine can do its magic.[7] If you're looking for an afternoon boost, head to the coffee shop between about 2 p.m. and 4 p.m., when cortisol levels dip again.

3. Soak up the morning sun.

If you feel sluggish in the morning, get as much sunlight as you can. The sun, unlike most lightbulbs, emits light that covers a wide swath of the color spectrum. When these extra wavelengths hit your eyes, they signal your brain to stop producing sleep hormones and start producing alertness hormones.

4. Schedule talk-therapy appointments for the morning.

Research in the emerging field of psychoneuroendocrinology has shown that therapy sessions may be most effective in the morning.[8] The reason goes back to cortisol. Yes, it's a stress hormone. But it also enhances learning. During therapy sessions in the morning, when cortisol levels are highest, patients are more focused and absorb advice more deeply.

2.

AFTERNOONS AND
COFFEE SPOONS

The Power of Breaks, the Promise of
Lunch, and the Case for a Modern Siesta

The afternoon knows what the morning never suspected.

—ROBERT FROST

C ome with me for a moment into the Hospital of Doom.

At this hospital, patients are three times more likely than at other hospitals to receive a potentially fatal dosage of anesthesia and considerably more likely to die within forty-eight hours of surgery. Gastroenterologists here find fewer polyps during colonoscopies than their more scrupulous colleagues, so cancerous growths go undetected. Internists are 26 percent more likely to prescribe unnecessary antibiotics for viral infections, thereby fueling the rise of drug-resistant superbugs. And throughout the facility, nurses and other caregivers are nearly 10 percent less likely to wash their hands before treating patients, increasing the probability that patients will contract an infection in the hospital they didn't have when they entered.

If I were a medical malpractice lawyer—and I'm thankful that I'm not—I'd hang out a shingle across the street from such a place. If I were a husband and parent—and I'm thankful that I am—I wouldn't let any member of my family walk through this hospital's doors. And if I were advising you on how to navigate your life—which, for better or worse, I'm doing in these pages—I'd offer the following counsel: Stay away.

The Hospital of Doom may not be a real name. But it is a real place. Everything I've described is what happens in modern medical centers during the afternoons compared with the mornings. Most hospitals and health care professionals do heroic work. Medical calamities are the exceptions rather than the norm. But afternoons can be a dangerous time to be a patient.

Something happens during the trough, which often emerges about seven hours after waking, that makes it far more perilous than any other time of the day. This chapter will examine why so many of us—from anesthesiologists to schoolchildren to the captain of the *Lusitania*—blunder in the afternoon. Then we'll look at some solutions for the problem—in particular, two simple remedies that can keep patients safer, boost students' test scores, and maybe even make the justice system fairer. Along the way, we'll learn why lunch (not breakfast) is the most important meal of the day, how to take a perfect nap, and why reviving a thousand-year-old practice may be just what we need today to boost individual productivity and corporate performance.

But first let's go into an actual hospital, where doom has been forestalled by lime-green laminated cards.

BERMUDA TRIANGLES AND PLASTIC RECTANGLES: THE POWER OF VIGILANCE BREAKS

It's a cloudy Tuesday afternoon in Ann Arbor, Michigan, and for the first (and probably only) time in my life, I'm wearing hospital greens and scrubbing in for surgery. Beside me is Dr. Kevin Tremper, an anesthesiologist and professor who is chairman of the University of Michigan Medical School's Department of Anesthesiology.

"Each year, we put 90,000 people to sleep and wake them up," he tells me. "We paralyze them and start cutting them open." Tremper oversees 150 physicians and another 150 medical residents who wield these magical powers. In 2010 he changed how they do their jobs.

Flat on the operating room table is a twenty-something man with a smashed jaw badly in need of repair. On a nearby wall is a large-screen television with the names of the five other people in hospital greens—nurses, physicians, a technician—who surround the table. At the top of the screen, in maize letters against a blue background, is the patient's name. The surgeon, an intense, wiry man in his thirties, is itching to begin. But before anybody does anything, as if this team were playing college basketball at the school's Crisler Center two miles away, they call a time-out.

Almost imperceptibly, each person takes one step backward. Then, looking at either the big screen or a wallet-size plastic card hanging from their waists, they introduce themselves to one another by first name and proceed through a nine-step "Pre-Induction Verification" checklist that ensures they've got the right patient, know his condition and any allergies, understand the medications the anesthesiologist will use, and have any special equipment they might need. When everyone is finished introducing themselves and all the questions are answered—the whole process takes about three minutes—the time-out ends and the young anesthesia resident cracks open supplies from sealed pouches to begin to put the patient, already partly sedated, fully to sleep. It's not easy. The patient's jaw is in such dreadful condition, the resident must intubate him through the nose instead of the mouth, which proves vexing. Tremper, who has the long, slender fingers of a pianist, steps in and steers the tube into the nasal cavity and down the patient's throat. Soon the patient is out, his vital signs are stable, and the surgery can begin.

Then the team steps back from the operating table once again.

Each person reviews the steps on the "Pre-Incision Time Out" card to make sure everyone is prepared. They regain their individual and collective focus. And only then does everyone step back to the operating table and the surgeon begins repairing the jaw.

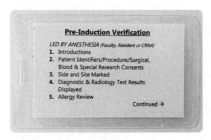
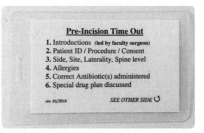

I call time-outs like these "vigilance breaks"—brief pauses before high-stakes encounters to review instructions and guard against error. Vigilance breaks have gone a long way in preventing the University of Michigan Medical Center from transmogrifying into the Hospital of Doom during the afternoon trough. Tremper says that in the time since he implemented these breaks, the quality of care has risen, complications have declined, and both doctors and patients are more at ease.

A fternoons are the Bermuda Triangles of our days. Across many domains, the trough represents a danger zone for productivity, ethics, and health. Anesthesia is one example. Researchers at Duke Medical Center reviewed about 90,000 surgeries at the hospital and identified what they called "anesthetic adverse events"—either mistakes anesthesiologists made, harm they caused to patients, or both. The trough was especially treacherous. Adverse events were significantly "more frequent for cases starting during the 3 p.m. and 4 p.m. hours." The probability of a problem at 9 a.m. was about 1 percent. At 4 p.m., 4.2 percent. In other words, the chance of something going awry while someone is delivering drugs to knock you unconscious was four times greater during the trough than during the peak. On actual harm (not only a slipup but also something that hurts the patient), the probability at 8 a.m. was 0.3 percent—three-tenths of one percent. But at 3 p.m., the probability was 1 percent—one in every one hundred cases, a threefold increase. Afternoon circadian lows, the researchers concluded, impair physician vigilance and "affect human performance of complex tasks such as those required in anesthesia care."[1]

Or consider colonoscopies. I've reached the age where prudence calls for submitting to this procedure to detect the presence or possibility of colon cancer. But now that I've read the research, I would never accept an appointment that wasn't before noon. For example, one oft-cited study of more than 1,000 colonoscopies found that endoscopists are less likely to detect polyps—small growths on the colon—as the day progresses. Every hour that passed resulted in a nearly 5 percent reduction in polyp detection. Some of the specific morning versus afternoon differences were stark. For instance, at 11 a.m., doctors found an average of more than 1.1 polyps in every exam. By 2 p.m., though, they were detecting barely half that number even though afternoon patients were no different from the morning ones.[2]

Look at those numbers and tell me when *you'd* schedule a colonoscopy.[3] What's more, other research has shown that doctors are significantly less likely even to fully complete a colonoscopy when they perform it in the afternoon.[4]

Basic health care also suffers when its practitioners sail into the day's Bermuda Triangle. Doctors, for example, are much more likely to prescribe antibiotics, including unnecessary ones, for acute respiratory infections in the afternoons than in the mornings.[5] As the cumulative effect of dealing with patient after patient saps doctors' decision-making resolve, it's far easier just to write the scrip than suss out whether the patient's symptoms suggest a bacterial infection, for which antibiotics might be appropriate, or a virus, for which they'd have no effect.

We expect important encounters with experienced professionals like physicians to turn on *who* is the patient and *what* is the problem. But many outcomes depend even more forcefully on *when* is the appointment.

What's going on is a decline in vigilance. In 2015, Hengchen Dai, Katherine Milkman, David Hoffman, and Bradley Staats led a massive study of handwashing at nearly three dozen U.S. hospitals. Using data from sanitizer dispensers equipped with radio frequency identification (RFID) to communicate with RFID chips on employee badges, researchers could monitor who washed their hands and who didn't. In all, they studied more than 4,000 caregivers (two-thirds of whom were nurses), who over the course of the research had nearly 14 million "hand hygiene opportunities." The results were not pretty. On average, these employees washed their hands less than half the time when they had an opportunity and a professional obligation to do so. Worse, the caregivers, most of whom began their shifts in the morning, were *even less* likely to sanitize their hands in the afternoons. This decline from the relative diligence of the mornings to the relative neglect of the afternoon was as great as 38 percent. That is, for every ten times they washed their hands in the morning, they did so only six times in the afternoon.[6]

The consequences are grave. "The decrease in hand hygiene compliance that we detected during a typical work shift would contribute to approximately 7,500 unnecessary infections per year at an annual cost of approximately $150 million across the 34 hospitals included in this study," the authors write. Spread this rate across annual hospital admissions in the United States, and the cost of the trough is massive: 600,000 unnecessary infections, $12.5 billion in added costs, and up to 35,000 unnecessary deaths.[7]

Afternoons can also be deadly beyond the white walls of a hospital. In the United Kingdom, sleep-related vehicle accidents peak twice during every twenty-four-hour period. One is between 2 a.m. and 6 a.m., the middle of the night. The other is between 2 p.m. and 4 p.m., the middle of the afternoon. Researchers have found the same pattern of traffic accidents in the United States, Israel, Finland, France, and other countries.[8]

One British survey got even more precise when it found that the typical worker reaches the most unproductive moment of the day at 2:55 p.m.[9] When we enter this region of the day, we often lose our bearings. In chapter 1, I briefly discussed the "morning morality effect," which found that people were more likely to be dishonest in the afternoon because most of us are "better able to resist opportunities to lie, cheat, steal and engage in other unethical behavior in the morning than in the afternoon."[10] This phenomenon depended in part on chronotype, with owls displaying a different pattern from larks or third birds. But in that study, evening types proved more ethical between midnight and 1:30 a.m., not during the afternoon.

Regardless of our chronotype, the afternoon can impair our professional and ethical judgment.

The good news is that vigilance breaks can loosen the trough's grip on our behavior. As the doctors at the University of Michigan demonstrate, inserting regular mandatory vigilance breaks into tasks helps us regain the focus needed to proceed with challenging work that must be done in the afternoon. Imagine if Captain Turner, who hadn't slept the night before his fateful decisions, had taken a brief vigilance break with other crew members to review how fast the *Lusitania* needed to travel and how best to calculate the ship's position in order to avoid U-boats.

This simple intervention is backed by heartening evidence. For instance, the largest health care system in the United States is the Veterans Health Administration, which operates about 170 hospitals across the country. In response to the persistence of medical errors (many of which occurred in afternoons), a team of physicians at the VA implemented a comprehensive training system across the hospitals (on which Michigan modeled its own efforts) that was built around the concept of more intentional and more frequent breaks, and featured such tools as "laminated checklist cards, whiteboards, paper forms, and wall mounted posters." One year after the training began, the surgical mortality rate (how often people died during or shortly after surgery) dropped 18 percent.[11]

Still, for most people, work doesn't involve paralyzing others and cutting them open—or other life-on-the-line responsibilities such as flying a twenty-seven-ton jet or guiding troops into battle. For the rest of us, another type of break offers a simple way to steer around the dangers of the trough. Call them "restorative breaks." And to understand them, let's leave the American Midwest and head to Scandinavia and the Middle East.

FROM THE SCHOOLHOUSE TO THE COURTHOUSE: THE POWER OF RESTORATIVE BREAKS

In chapter 1 we learned about some curious results on Denmark's national standardized exams. Danish schoolchildren who take the tests in the afternoon score significantly worse than those who take the exams earlier in the day. To a school principal or education policy maker, the response seems obvious: Whatever it takes, move all the tests to the morning. However, the researchers also discovered another remedy, one with applications beyond schools and tests, that is remarkably easy to explain and implement.

When the Danish students had a twenty- to thirty-minute break "to eat, play, and chat" before a test, their scores did not decline. In fact, they *increased*. As the researchers note, "A break causes an improvement that is larger than the hourly deterioration."[12] That is, scores go down after noon. But scores go up by a higher amount after breaks.

Taking a test in the afternoon without a break produces scores that are equivalent to spending less time in school each year and having parents with lower incomes and less education. But taking the same test after a twenty- to thirty-minute break leads to scores that are equivalent to students spending three *additional* weeks in the classroom and having somewhat *wealthier* and *better-educated* parents. And the benefits were the greatest for the lowest-performing students.

Unfortunately, Danish schools, like many around the world, offer only two breaks each day. Worse, legions of school systems are cutting back on recess and other restorative pauses for students in the name of rigor and—get ready for the irony—higher test scores. But as Harvard's Francesca Gino, one of the study's authors, puts it, "If there were a break after every hour, test scores would actually improve over the course of the day."[13]

Many younger students underperform during the trough, which risks both providing teachers with an inaccurate sense of their progress and prompting administrators to attribute to *what* and *how* students are learning something that is really an issue of *when* they're taking a test. "We believe these results to have two important policy implications," say the researchers who studied the Danish experience. "[F]irst, cognitive fatigue should be taken into consideration when deciding on the length of the school day and the frequency and duration of breaks. Our results show that longer school days can be justified, if they include an appropriate number of breaks. Second, school accountability systems should control for the influence of external factors on test scores . . . a more straightforward approach would be to plan tests as closely after breaks as possible."[14]

Perhaps it makes sense that a cup of apple juice and a few minutes to run around works wonders for eight-year-olds solving arithmetic problems. But restorative breaks have a similar power for adults with weightier responsibilities.

In Israel, two judicial boards process about 40 percent of the country's parole requests. At their helm are individual judges whose job is to hear prisoners' cases one after another and make decisions about their fate. Should this prisoner be released because she's served enough time on her sentence and shown sufficient signs of rehabilitation? Should that one, already granted parole, now be permitted to move about without his tracking device?

Judges aspire to be rational, deliberative, and wise, to mete out justice based on the facts and the law. But judges are also human beings subject to the same daily rhythms as the rest of us. Their black robes don't shelter them from the trough. In 2011 three social scientists (two Israelis and one American) used data from these two parole boards to examine judicial decision-making. They found that, in general, judges were more likely to issue a favorable ruling—granting the prisoner parole or allowing him to remove an ankle monitor—in the morning than in the afternoon. (The study controlled for the type of prisoner, the severity of the offense, and other factors.) But the pattern of decision-making was more complicated, and more intriguing, than a simple a.m./p.m. divide.

The following chart shows what happened. Early in the day, judges ruled in favor of prisoners about 65 percent of the time. But as the morning wore on, that rate declined. And by late morning, their favorable rulings dropped to nearly zero. So a prisoner slotted for a 9 a.m. hearing was likely to get parole while one slotted for 11:45 a.m. had essentially no chance at all—regardless of the facts of the case. Put another way, since the default decision on boards is typically not to grant parole, judges deviated from the status quo during some hours and reinforced it during others.

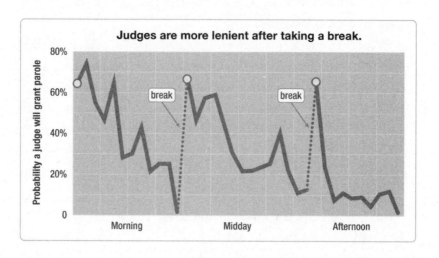

But look what happens after the judges take a break. Immediately after that first break, for lunch, they become more forgiving—more willing to deviate from the default—only to sink into a more hard-line attitude after a few hours. But, as happened with the Danish schoolchildren, look what occurs when those judges then get a second break—a midafternoon restorative pause to drink some juice or play on the judicial jungle gym. They return to the same rate of favorable decisions they displayed first thing in the morning.

Ponder the consequences: If you happen to appear before a parole board just before a break rather than just after one, you'll likely spend a few more years in jail—not because of the facts of the case but because of the time of day. The researchers say they cannot identify precisely what's driving this phenomenon. It could be that eating restored judges' glucose levels and replenished their mental reserves. It could be that a little time away from the bench lifted their mood. It could be that the judges were tired and that rest reduced their fatigue. (Another study of U.S. federal courts found that on the Mondays after the switch to Daylight Saving Time, when people on average lose roughly forty minutes of sleep, judges rendered prison sentences that were about 5 percent longer than the ones they handed down on typical Mondays.[15])

Whatever the explanation, a factor that should have been extraneous to judicial decision-making and irrelevant to justice itself—whether and when a judge took a break—was critical in deciding whether someone would go free or remain behind bars. And the wider phenomenon—that breaks can often mitigate the trough—likely applies "in other important sequential decisions or judgments, such as legislative decisions . . . financial decisions, and university admissions decisions."[16]

So if the trough is the poison and restorative breaks are the antidote, what should those breaks look like? There's no single answer, but science offers five guiding principles.

1. Something beats nothing.

One problem with afternoons is that if we stick with a task too long, we lose sight of the goal we're trying to achieve, a process known as "habituation." Short breaks from a task can prevent habituation, help us maintain focus, and reactivate our commitment to a goal.[17] And frequent short breaks are more effective than occasional ones.[18] DeskTime, a company that makes productivity-tracking software, says that "what the most productive 10% of our users have in common is their ability to take effective breaks." Specifically, after analyzing its own data, DeskTime claims to have discovered a golden ratio of work and rest. High performers, its research concludes, work for fifty-two minutes and then break for seventeen minutes. DeskTime never published the data in a peer-reviewed journal, so your mileage may vary. But the evidence is overwhelming that short breaks are effective—and deliver considerable bang for their limited buck. Even "micro-breaks" can be helpful.[19]

2. Moving beats stationary.

Sitting, we've been told, is the new smoking—a clear and present danger to our health. But it also leaves us more susceptible to the dangers of the trough, which is why simply standing up and walking around for five minutes every hour during the workday can be potent. One study showed that hourly five-minute walking breaks boosted energy levels, sharpened focus, and "improved mood throughout the day and reduced feelings of fatigue in the late afternoon." These "microbursts of activity," as the researchers call them, were also more effective than a single thirty-minute walking break—so much so that the researchers suggest that organizations "introduce physically active breaks during the workday routine."[20] Regular short walking breaks in the workplace also increase motivation and concentration and enhance creativity.[21]

3. Social beats solo.

Time alone can be replenishing, especially for us introverts. But much of the research on restorative breaks points toward the greater power of being with others, particularly when we're free to choose with whom we spend the time. In high-stress occupations like nursing, social and collective rest breaks not only minimize physical strain and cut down on medical errors, they also reduce turnover; nurses who take these sorts of breaks are more likely to stay at their jobs.[22] Likewise, research in South Korean workplaces shows that social breaks—talking with coworkers about something other than work—are more effective at reducing stress and improving mood than either cognitive breaks (answering e-mail) or nutrition breaks (getting a snack).[23]

4. Outside beats inside.

Nature breaks may replenish us the most.[24] Being close to trees, plants, rivers, and streams is a powerful mental restorative, one whose potency most of us don't appreciate.[25] For example, people who take short walks outdoors return with better moods and greater replenishment than people who walk indoors. What's more, while people predicted they'd be happier being outside, they underestimated *how much* happier.[26] Taking a few minutes to be in nature is better than spending those minutes in a building. Looking out a window into nature is a better micro-break than looking at a wall or your cubicle. Even taking a break indoors amid plants is better than doing so in a green-free zone.

5. Fully detached beats semidetached.

By now, it's well known that 99 percent of us cannot multitask. Yet, when we take a break, we often try to combine it with another cognitively demanding activity—perhaps checking our text messages or talking to a colleague about a work issue. That's a mistake. In the same South Korean study mentioned earlier, relaxation breaks (stretching or daydreaming) eased stress and boosted mood in a way that multitasking breaks did not.[27] Tech-free breaks also "increase vigor and reduce emotional exhaustion."[28] Or, as other researchers put it, "Psychological detachment from work, in addition to physical detachment, is crucial, as continuing to think about job demands during breaks may result in strain."[29]

So if you're looking for the Platonic ideal of a restorative break, the perfect combination of scarf, hat, and gloves to insulate yourself from the cold breath of the afternoon, consider a short walk outside with a friend during which you discuss something other than work.

Vigilance breaks and restorative breaks offer us a chance to recharge and replenish, whether we're performing surgery or proofreading advertising copy. But two other respites are also worth considering. Both were once sturdy features of professional and personal life only to be dismissed more recently as soft, frivolous, and antithetical to the head-down, laptop-up, inbox-zero ethos of the twenty-first century. Now both are poised for a comeback.

THE MOST IMPORTANT MEAL OF THE DAY

After you woke up this morning, some time before you began a day of filing reports, making deliveries, or chasing children, you probably ate breakfast. You might not have settled in for a full, proper meal, but I'll bet you broke the nighttime fast with something—a piece of toast maybe or a little yogurt, perhaps washed down with coffee or tea. Breakfast fortifies our bodies and fuels our brains. It's also a guardrail for our metabolism; eating breakfast restrains us from gorging the rest of the day, which keeps our weight down and our cholesterol in check. These truths are so self-evident, these benefits so manifest, that the principle has become a nutritional catechism. Say it with me: Breakfast is the most important meal of the day.

As a devout breakfast eater, I endorse this principle. But as some-one paid to muck around in scientific journals, I've grown skeptical. Most of the research showing the salvation of a morning meal and the sin of missing it are observational studies rather than randomized controlled experiments. Researchers follow people around, watching what they do, but they don't compare them to a control group.[30] That means their findings show correlation (people who eat break-fast might well be healthy) but not necessarily causation (maybe people who are already healthy are just more likely to eat breakfast). When scholars have applied more rigorous scientific methods, break-fast's benefits have been much more difficult to detect.

"A recommendation to eat or skip breakfast . . . contrary to widely espoused views . . . had no discernable effect on weight loss," says one.[31] "The belief (in breakfast) . . . exceeds the strength of scientific evidence," says another.[32] Layer in the fact that several studies show-ing the virtues of breakfast were funded by industry groups and the skepticism deepens.

Should we all eat breakfast? The conventional view is a flaky and delicious yes. But as a leading British nutritionist and statistician says, "[T]he current state of scientific evidence means that, unfortu-nately, the simple answer is: I don't know."[33]

So eat breakfast if you'd like. Or skip it if you'd prefer. But if you're concerned about the perils of the afternoon, start taking more seriously the often-maligned and easily dismissed meal called lunch. ("Lunch is for wimps," 1980s cinematic supervillain Gordon Gekko famously declared.) By one estimation, 62 percent of American office workers wolf down lunch in the same spot where they work all day. These dismal scenes—smartphone in one hand, soggy sandwich in the other, despair wafting from the cubicle—even have a name: the sad desk lunch. And that name has given rise to a small online movement in which people post photographs of their oh-so-pathetic midday meals.[34] But it's time we paid more attention to lunch, because social scientists are discovering that it's far more important to our performance than we realize.

For example, a 2016 study looked at more than eight hundred workers (mostly in information technology, education, and media) from eleven different organizations, some of whom regularly took lunch breaks away from their desks and some of whom did not. The non–desk lunchers were better able to contend with workplace stress and showed less exhaustion and greater vigor not just during the remainder of the day but also a full one year later.

"Lunch breaks," the researchers say, "offer an important recovery setting to promote occupational health and well-being"—particularly for "employees in cognitively or emotionally demanding jobs."[35] For groups that require high levels of cooperation—say, firefighters—eating together also enhances team performance.[36]

Not just any lunch will do, however. The most powerful lunch breaks have two key ingredients—autonomy and detachment. Autonomy—exercising some control over what you do, how you do it, when you do it, and whom you do it with—is critical for high performance, especially on complex tasks. But it's equally crucial when we take breaks from complex tasks. "The extent to which employees can determine how they utilize their lunch breaks may be just as important as what employees do during their lunch," says one set of researchers.[37]

Detachment—both psychological and physical—is also critical. Staying focused on work during lunch, or even using one's phone for social media, can intensify fatigue, according to multiple studies, but shifting one's focus away from the office has the opposite effect. Longer lunch breaks and lunch breaks away from the office can be prophylactic against afternoon peril. Some of these researchers suggest that "organizations could promote lunchtime recovery by giving options to spend lunch breaks in different ways that enable detachment, such as spending a break in a non-work environment or offering a space for relaxing activities."[38] Ever so slowly, organizations are responding. For instance, in Toronto, CBRE, the large commercial real estate firm, has banned desk lunches in the hope that employees will take a proper lunch break.[39]

Given this evidence, as well as the dangers of the trough, it's becoming ever clearer that we must revise some oft-repeated advice. Say it with me now, brothers and sisters: Lunch is the most important meal of the day.

SLEEPING ON THE JOB

I hate naps. Maybe I enjoyed them when I was a kid. But from the age of five onward, I've considered them the behavioral equivalent of sippy cups—fine for toddlers, pathetic for grown-ups. It's not that I've never napped as an adult. I have—sometimes intentionally, most times inadvertently. But when I've awoken from these slumbers, I usually feel woozy, wobbly, and befuddled—shrouded in a haze of grogginess and enveloped in a larger cloud of shame. To me, naps are less an element of self-care than a source of self-loathing. They are a sign of personal failure and moral weakness.

But I've recently changed my mind. And in response, I've changed my ways. Done right, naps can be a shrewd response to the trough and a valuable break. Naps, research shows, confer two key benefits: They improve cognitive performance and they boost mental and physical health.

In many ways, naps are Zambonis for our brains. They smooth out the nicks, scuffs, and scratches a typical day has left on our mental ice. One well-known NASA study, for instance, found that pilots who napped for up to forty minutes subsequently showed a 34 percent improvement in reaction time and a twofold increase in alertness.[40] The same benefit redounds to air traffic controllers: After a short nap, their alertness sharpens and their performance climbs.[41] Italian police officers who took naps immediately before their afternoon and evening shifts had 48 percent fewer traffic accidents than those who didn't nap.[42]

However, the returns from napping extend beyond vigilance. An afternoon nap expands the brain's capacity to learn, according to a University of California–Berkeley study. Nappers easily outperformed non-nappers on their ability to retain information.[43] In another experiment, nappers were twice as likely to solve a complex problem than people who hadn't napped or who had spent the time in other activities.[44] Napping boosts short-term memory as well as associative memory, the type of memory that allows us to match a face to a name.[45] The overall benefits of napping to our brainpower are massive, especially the older we get.[46] As one academic overview of the napping literature explains, "Even for individuals who generally get the sleep they need on a nightly basis, napping may lead to considerable benefits in terms of mood, alertness and cognitive performance. . . [It] is particularly beneficial to performance on tasks, such as addition, logical reasoning, reaction time, and symbol recognition."[47] Napping even increases "flow," that profoundly powerful source of engagement and creativity.[48]

Naps also improve our overall health. A large study in Greece, which followed more than 23,000 people over six years, found that, controlling for other risk factors, people who napped were as much as 37 percent less likely as others to die from heart disease, "an effect of the same order of magnitude as taking an aspirin or exercising every day."[49] Napping strengthens our immune system.[50] And one British study found that simply anticipating a nap can reduce blood pressure.[51]

Yet, even after absorbing this evidence, I remained a nap skeptic. One reason I so disliked naps is that I woke up from them feeling as if someone had injected my bloodstream with oatmeal and replaced my brain with oily rags. Then I discovered something crucial: I was doing it wrong.

While naps between thirty and ninety minutes can produce some long-term benefits, they come with steep costs. The ideal naps—those that combine effectiveness with efficiency—are far shorter, usually between ten and twenty minutes. For instance, an Australian study published in the journal *Sleep* found that five-minute naps did little to reduce fatigue, increase vigor, or sharpen thinking. But ten-minute naps had positive effects that lasted nearly three hours. Slightly longer naps were also effective. But once the nap lasted beyond about the twenty-minute mark, our body and brain began to pay a price.[52] That price is known as "sleep inertia"—the confused, boggy feeling I typically had upon waking. Having to recover from sleep inertia—all that time spent splashing water on my face, shaking my upper body like a soaked golden retriever, and searching desk drawers for candy to get some sugar into my system—subtracts from the nap's benefits, as this chart makes clear.

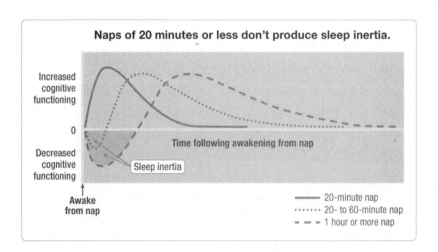

With brief ten- to twenty-minute naps, the effect on cognitive functioning is positive from the moment of awakening. But with slightly longer snoozes, the napper begins in negative territory—that's sleep inertia—and must dig herself out. And with naps of more than an hour, cognitive functioning drops for even longer before it reaches a prenap state and eventually turns positive.[53] In general, concludes one analysis of about twenty years of napping research, healthy adults "should ideally nap for approximately 10 to 20 minutes." Such brief naps "are ideal for workplace settings where performance immediately upon awakening is usually required."[54]

But I also learned I was making another mistake. Not only was I taking the wrong kind of nap, I was also failing to use a potent (and legal) drug that can enhance a short nap's benefits. To paraphrase T. S. Eliot, we should measure out our naps in coffee spoons.

One study makes this case. The experimenters divided participants into three groups and gave them all a thirty-minute midafternoon break before sitting them at a driving simulator. One group received a placebo pill. The second received two hundred milligrams of caffeine. The third received that same two hundred milligrams of caffeine and then took a brief nap. When it came time to perform, the caffeine-only group outperformed the placebo group. But the group that ingested caffeine and then had a nap easily bested them both.[55] Since caffeine takes about twenty-five minutes to enter the bloodstream, they were getting a secondary boost from the drug by the time their naps were ending. Other researchers have found the same results—that caffeine, usually in the form of coffee, followed by a nap of ten to twenty minutes, is the ideal technique for staving off sleepiness and increasing performance.[56]

As for me, after a few months of experimenting with twenty-minute afternoon naps, I've converted. I've gone from nap detractor to nap devotee, from someone ashamed to nap to someone who relishes the coffee-then-nap combination known as the "nappuccino."*

---

* See this chapter's Time Hacker's Handbook for nappuccino instructions and how to take a perfect nap.

THE CASE FOR A MODERN SIESTA

A decade ago, the government of Spain took a step that seemed distinctly un-Spanish: It officially eliminated the siesta. For centuries, Spaniards had enjoyed an afternoon respite, often returning home to eat a meal with their family and even snag a quick sleep. But Spain, its economy sluggish, was determined to reckon with twenty-first-century realities. With two parents working, and globalization tightening competition worldwide, this lovely practice was stifling Spanish prosperity.[57] Americans applauded the move. Spain was finally treating work with sufficient, and sufficiently puritanical, seriousness. At last, Old Europe was becoming modern.

But what if this now-eliminated practice was actually a stroke of genius, less an indulgent relic than a productivity-boosting innovation?

In this chapter, we've seen that breaks matter—that even little ones can make a big difference. Vigilance breaks prevent deadly mistakes. Restorative breaks enhance performance. Lunches and naps help us elude the trough and get more and better work done in the afternoon. A growing body of science makes it clear: Breaks are not a sign of sloth but a sign of strength.

So instead of celebrating the death of the siesta, perhaps we should consider resurrecting it—though in a form more appropriate for contemporary work life. "Siesta" derives from the Latin *hora sexta*, which means "sixth hour." It was during the sixth hour after dawn that these breaks usually began. In ancient times, when most people worked outside and indoor air-conditioning was still a few thousand years away, escaping the midday sun was a physical imperative. Today, escaping the midafternoon trough is a psychological imperative.

Likewise, the Koran, which a thousand years ago identified sleep stages that align with modern science, also calls for a midday break.

It "is a deeply embedded practice in the Muslim culture, and it takes a religious dimension (*Sunnah*) for some Muslims," says one scholar.[58]

Maybe breaks can become a deeply embedded organizational practice with a scientific and secular dimension.

A modern siesta does not mean giving everyone two or three hours off in the middle of the day. That's not realistic. But it does mean treating breaks as an essential component of an organization's architecture—understanding breaks not as a softhearted concession but as a hardheaded solution. It means discouraging sad desk lunches and encouraging people to go outside for forty-five minutes. It means protecting and extending recess for schoolchildren rather than eliminating it. It might even mean following the lead of Ben & Jerry's, Zappos, Uber, and Nike, all of which have created napping spaces for employees in their offices. (Alas, it probably does not mean legislating a one-hour break each week for employees to go home and have sex, as one Swedish town has proposed.[59])

Most of all, it means changing the way we think about what we do and how we can do it effectively. Until about ten years ago, we admired those who could survive on only four hours of sleep and those stalwarts who worked through the night. They were heroes, people whose fierce devotion and commitment revealed everyone else's fecklessness and frailty. Then, as sleep science reached the mainstream, we began to change our attitude. That sleepless guy wasn't a hero. He was a fool. He was likely doing subpar work and maybe hurting the rest of us because of his poor choices.

Breaks are now where sleep was then. Skipping lunch was once a badge of honor and taking a nap a mark of shame. No more. The science of timing now affirms what the Old World already understood: We should give ourselves a break.

Time Hacker's Handbook

• CHAPTER 2 •

MAKE A BREAK LIST

You probably have a to-do list. Now it's time to create a "break list," give it equal attention, and treat it with equal respect. Each day, alongside your list of tasks to complete, meetings to attend, and deadlines to hit, make a list of the breaks you're going to take.

Start by trying three breaks per day. List when you're going to take those breaks, how long they're going to last, and what you're going to do in each. Even better, put the breaks into your phone or computer calendar so one of those annoying pings will remind you. Remember: What gets scheduled gets done.

HOW TO TAKE A PERFECT NAP

As I explained, I've discovered the errors in my napping ways and have learned the secrets of a perfect nap. Just follow these five steps:

1. **Find your afternoon trough time.** The Mayo Clinic says that the best time for a nap is between 2 p.m. and 3 p.m.[1] But if you want to be more precise, take a week to chart your afternoon mood and energy levels, as described on pages 40–43. You'll likely see a consistent block of time when things begin to go south, which for many people is about seven hours after waking. This is your optimal nap time.

2. **Create a peaceful environment.** Turn off your phone notifications. If you've got a door, close it. If you've got a couch, use it. To insulate yourself from sound and light, try earplugs or headphones and an eye mask.

3. **Down a cup of coffee.** Seriously. The most efficient nap is the nappuccino. The caffeine won't fully engage in your bloodstream for about twenty-five minutes, so drink up right before you lie down. If you're not a coffee drinker, search online for an alternative drink that provides about two hundred milligrams of caffeine. (If you avoid caffeine, skip this step. Also reconsider your life choices.)

4. **Set a timer on your phone for twenty-five minutes.** If you nap for more than about a half hour, sleep inertia takes over and you need extra time to recover. If you nap for less than five minutes, you don't get much benefit. But naps between ten and twenty minutes measurably boost alertness and mental function, and don't leave you feeling even sleepier than you were before. Since it takes most people about seven minutes to nod off, the twenty-five-minute countdown clock is ideal. And, of course, when you wake up, the caffeine is beginning to kick in.

5. **Repeat consistently.** There's some evidence that habitual nappers get more from their naps than infrequent nappers. So if you have the flexibility to take a regular afternoon nap, consider making it a common ritual. If you don't have the flexibility, then pick days when you're really dipping—when you haven't gotten enough sleep the night before or the stress and demands of the day are weightier than usual. You'll feel a difference.

FIVE KINDS OF RESTORATIVE BREAKS: A MENU

You now understand the science of breaks and why they're so effective in both combatting the trough and boosting your mood and performance. You've even got a break list ready to go. But what sort of break should you take? There's no right answer. Just choose one from the following menu or combine a few, see how they go, and design the breaks that work best for you:

1. **Micro-breaks**—A replenishing break need not be lengthy. Even breaks that last a minute or less—what researchers call "micro-breaks"—can pay dividends.[2] Consider these:

   The 20–20–20 rule—Before you begin a task, set a timer. Then, every twenty minutes, look at something twenty feet away for twenty seconds. If you're working at a computer, this micro-break will rest your eyes and improve your posture, both of which can fight fatigue.

   Hydrate—You might already have a water bottle. Get a much smaller one. When it runs out—and of course it will, because of its size—walk to the water fountain and refill it. It's a threefer: hydration, motion, and restoration.

   Wiggle your body to reset your mind—One of the simplest breaks of all: Stand up for sixty seconds, shake your arms and legs, flex your muscles, rotate your core, sit back down.

2. **Moving breaks**—Most of us sit too much and move too little. So build more movement into your breaks. Some options:

   Take a five-minute walk every hour—As we have learned, five-minute walk breaks are powerful. They're feasible for most people. And they're especially useful during the trough.

Office yoga—You can do yoga poses right at your desk—chair rolls, wrist releases, forward folds—to relieve tension in your neck and lower back, limber up your typing fingers, and relax your shoulders. This may not be for everyone, but anyone can give it a try. Just stick "office yoga" into a search engine.

Push-ups—Yeah, push-ups. Do two a day for a week. Then four a day for the next week and six a day a week after that. You'll boost your heart rate, shake off cognitive cobwebs, and maybe get a little stronger.

3. **Nature break**—This might sound tree hugger-y, but study after study has shown the replenishing effects of nature. What's more, people consistently underestimate how much better nature makes them feel. Choose:

Walk outside—If you've got a few minutes and are near a local park, take a lap through it. If you work at home and have a dog, take Fido for a walk.

Go outside—If there are trees and a bench behind your building, sit there instead of inside.

Pretend you're outside—If the best you can do is look at some indoor plants or the trees outside your window—well, research suggests that will help, too.

4. **Social break**—Don't go it alone. At least not always. Social breaks are effective, especially when *you* decide the who and how. A few ideas:

Reach out and touch somebody—Call someone you haven't talked to for a while and just catch up for five or ten minutes. Reawakening these "dormant ties" is also a great way to strengthen your network.[3] Or use the moment to say thank you—via a note, an e-mail, or a quick visit—to someone who's helped you. Gratitude—with its mighty combination of meaning and social connection—is a mighty restorative.[4]

Schedule it—Plan a regular walk or visit to a coffee joint or weekly gossip session with colleagues you like. A fringe benefit of social breaks is that you're more likely to take one if someone else is counting on you. Or go Swedish and try what Swedes call a *fika*—a full-fledged coffee break that is the supposed key to Sweden's high levels of employee satisfaction and productivity.[5]

Don't schedule it—If your schedule is too tight for something regular, buy someone a coffee one day this week. Bring it to her. Sit and talk about something other than work for five minutes.

5. **Mental gear-shifting break**—Our brains suffer fatigue just as much as our bodies do—and that's a big factor in the trough. Give your brain a break by trying these:

Meditate—Meditation is one of the most effective breaks—and micro-breaks—of all.[6] Check out material from UCLA (http://marc.ucla.edu/mindful-meditations), which offers guided meditations as short as three minutes.

Controlled breathing—Have forty-five seconds? Then, as the *New York Times* explains: "Take a deep breath, expanding your belly. Pause. Exhale slowly to the count of five. Repeat four times."[7] It's called controlled breathing, and it can tamp your stress hormones, sharpen your thinking, and maybe even boost your immune system—all in under a minute.

Lighten up—Listen to a comedy podcast. Read a joke book. If you can find a little privacy, put on your headphones and jam out for a song or two. There's even evidence from one study on the replenishing effects of watching dog videos.[8] (No, really.)

CREATE YOUR OWN TIME-OUT AND TROUGH CHECKLIST

Sometimes it's not possible to pull completely away from an important task or project to take a restorative break. When you and your team need to plow forward and get a job done even if you're in the trough, that's when it's time for a vigilance break that combines a time-out with a checklist.

Here's how to plan it:

If you have a task or project that will need your continued vigilance and focus even through the trough, find a stage in the middle of that task to schedule a time-out. Plan for that time-out by creating a trough checklist modeled on the lime-green cards used at the University of Michigan Medical Center.

For example, suppose your team needs to get a major proposal out the door by 5 p.m. today. No one can afford to step outside and take a walk. Instead, schedule a time-out two hours before the deadline for everyone to gather. Your checklist might read:

1. Everyone stops what they are doing, takes a step backward, and draws a deep breath.
2. Each team member takes thirty seconds to report on their progress.
3. Each team member takes thirty seconds to describe their next step.
4. Each team member answers this question: What are we missing?
5. Assign who will address the missing pieces.
6. Schedule another time-out, if necessary.

PAUSE LIKE A PRO

Anders Ericsson is "the world expert on world experts."[9] A psychologist who studies extraordinary performers, Ericsson found that elite performers have something in common: They're really good at taking breaks.

Most expert musicians and athletes begin practicing in earnest around nine o'clock in the morning, hit their peak during the late morning, break in the afternoon, and then practice for a few more hours in the evening. For example, the practice pattern of the most accomplished violinists looks like this:

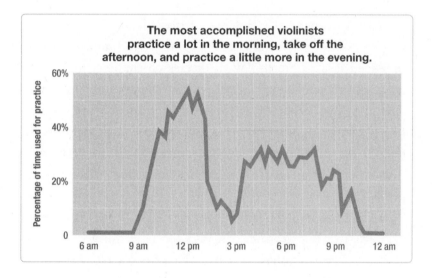

The most accomplished violinists practice a lot in the morning, take off the afternoon, and practice a little more in the evening.

Recognize that shape?

In Ericsson's study, one factor that distinguished the best from the rest is that they took *complete* breaks during the afternoon (many even napped as part of their routine), whereas nonexperts were less rigorous about pauses. We might think that superstars power straight through the day for hours on end. In fact, they practice with intense focus for forty-five- to ninety-minute bursts, then take meaningful restorative breaks.

You can do the same. Pause like a pro and you might become one.

GIVE KIDS A BREAK: THE HARDHEADED CASE FOR RECESS

Schools are getting tough. Especially in the United States, they are embracing high-stakes testing, strict evaluations of teachers, and a tough-minded approach to accountability. Some of these measures make sense, but the war on weakness has produced a major casualty: recess.

Some 40 percent of U.S. schools (particularly schools with large numbers of low-income students of color) have eliminated recess or combined it with lunch.[10] With futures on the line, the thinking goes, schools can't afford the frivolity of playtime. For example, in 2016 the New Jersey legislature passed a bipartisan bill requiring merely twenty minutes of recess each day for grades kindergarten to 5 in the state's schools. But Governor Chris Christie vetoed it, explaining in language reminiscent of a schoolyard, "That was a stupid bill."[11]

All this supposed toughness is wrongheaded. Breaks and recess are not deviations from learning. They are *part* of learning.

Years of research show that recess benefits schoolchildren in just about every realm of their young lives. Kids who have recess work harder, fidget less, and focus more intently.[12] They often earn better grades than those with fewer recesses.[13] They develop better social skills, show greater empathy, and cause fewer disruptions.[14] They even eat healthier food.[15] In short, if you want kids to flourish, let them leave the classroom.

What can schools do to take advantage of recess? Here are six pieces of guidance:

1. **Schedule recess before lunch.** A fifteen-minute break suffices, and it's the most helpful time for kids' concentration. It also makes them hungrier, so they eat better at lunch.
2. **Go minimalist.** Recess doesn't have to be tightly structured, nor does it need specialized equipment. Kids derive benefits from negotiating their own rules.
3. **Don't skimp.** In Finland, a nation with one of the world's highest-performing school systems, students get a fifteen-minute break every hour. Some U.S. schools—for instance, Eagle Mountain Elementary School in Fort Worth, Texas—have followed the Finnish lead and increased learning by offering four recesses each day for younger students.[16]
4. **Give teachers a break.** Schedule recesses in shifts so teachers can alternate monitoring duties with breaks for themselves.
5. **Don't replace physical education.** Structured PE is a separate part of learning, not a substitute for recess.
6. **Every kid, every day.** Avoid using the denial of recess as a punishment. It's essential to every kid's success, even those who slip up. Ensure that every student gets recess every school day.

PART TWO. BEGINNINGS, ENDINGS, AND IN BETWEEN

3.

BEGINNINGS

Starting Right, Starting Again, and Starting Together

Todo es comenzar á ser venturoso.
(To be lucky at the beginning is everything.)

—MIGUEL DE CERVANTES, *Don Quixote*

Every Friday, the U.S. Centers for Disease Control and Prevention, the government agency charged with protecting American citizens from health threats, issues a publication called the *Morbidity and Mortality Weekly Report*. Although the *MMWR* is written in the etherized prose of many government documents, its contents can be as terrifying as a Stephen King novel. Each edition offers a fresh menu of menaces—not just marquee diseases such as Ebola, hepatitis, and West Nile virus but also lesser-known dangers such as human pneumonic plague, rabies in dogs imported from Egypt, and elevated carbon monoxide levels in indoor skating rinks.

The full contents of the *MMWR* for the first week of August 2015 were no more alarming than usual. But for American parents, the five-page lead article was chilling. The CDC had identified a disease endangering roughly 26 million American teenagers. This threat, the report showed, was pelting young people with a hailstorm of dangers:

- Weight gain and a greater likelihood of being overweight
- Symptoms of clinical depression
- Lower academic performance
- A higher propensity "to engage in unhealthy risk behaviors such as drinking, smoking tobacco, and using illicit drugs"[1]

Meanwhile, researchers at Yale University were busy identifying a threat to some of these beleaguered teenagers' older brothers and sisters. This hazard wasn't imperiling their physical or emotional health—at least not yet—but it was gnawing at their livelihoods. These men and women in their mid to late twenties were stalled. Even though they had graduated from college, they were earning less than they had expected with a bachelor's degree and significantly less than people who'd graduated just a few years earlier. And this was no short-term problem. They would suffer from reduced wages for a decade, maybe longer. Nor was this cluster of twenty-somethings alone. Some of their parents, who had graduated college in the early 1980s, had suffered from the same malady and were still trying to shake off its residue.

What had gone so wrong for so many?

The full answer is a complex blend of biology, psychology, and public policy. But the core explanation is simple: These people were suffering because they had gotten off to a bad start.

In the case of those teenagers, they were starting the school day far too early—and that was jeopardizing their ability to learn. In the case of those twenty-somethings, and even some of their mothers and fathers, they had begun their careers, through no fault of their own, during a recession—and that was depressing their earnings years and years beyond their first job.

Faced with problems as vexing as underperforming teenagers or flattened wages, we often search for solutions in the realm of *what*. What are people doing wrong? What can they do better? What can others do to help? But, more frequently than we realize, the most potent answers lurk in the realm of *when*. In particular, when we begin—the school day, a career—can play an outsize role in our personal and collective fortunes. For teenagers, beginning the school day before 8:30 a.m. can impair their health and hobble their grades, which, in turn, can limit their options and alter the trajectory of their lives. For somewhat older people, beginning a career in a weak economy can restrict opportunities and reduce earning power well into adulthood. Beginnings have a far greater impact than most of us understand. Beginnings, in fact, can matter to the end.

Although we can't always determine when we start, we can exert some influence on beginnings—and considerable influence on the consequences of less than ideal ones. The recipe is straightforward. In most endeavors, we should be awake to the power of beginnings and aim to make a strong start. If that fails, we can try to make a fresh start. And if the beginning is beyond our control, we can enlist others to attempt a group start. These are the three principles of successful beginnings: Start right. Start again. Start together.

STARTING RIGHT

In high school, I studied French for four years. I don't remember much of what I learned, but one aspect of French class that I do recall might explain some of my deficiencies. Mademoiselle Inglis's class met first period—around 7:55 a.m., I think. She would usually warm us up by posing the question that French teachers—from the European language academies of the seventeenth century to my own central Ohio public school in the 1980s—have always asked their students: *Comment allez-vous?* "How are you?"

In Mlle. Inglis's class, every answer from every student on every morning was the same: *Je suis fatigué.* "I'm tired." Richard was *fatigué.* Lori was *fatiguée.* I myself was frequently *très fatigué.* To a French-speaking visitor, my twenty-six classmates and I probably sounded as if we were suffering from a bizarre form of group narcolepsy. *Quelle horreur! Tout le monde est fatigué!*

But the real explanation is less exotic. We were all just teenagers trying to use our brains before eight o'clock in the morning.

As I explained in chapter 1, young people begin undergoing the most profound change in chronobiology of their lifetimes around puberty. They fall asleep later in the evening and, left to their own biological imperatives, wake up later in the morning—a period of peak owliness that stretches into their early twenties.

Yet most secondary schools around the world force these extreme owls into schedules designed for chirpy seven-year-old larks. The result is that teenage students sacrifice sleep and suffer the consequences. "Adolescents who get less sleep than they need are at higher risk for depression, suicide, substance abuse and car crashes," according to the journal *Pediatrics.* "Evidence also links short sleep duration with obesity and a weakened immune system."[2] While younger students score higher on standardized tests scheduled in the morning, teenagers do better later in the day. Early start times correlate strongly with worse grades and lower test scores, especially in math and language.[3] Indeed, a study from McGill University and the Douglas Mental Health University Institute, both in Montreal, found that the amount and quality of sleep explained a sizable portion of the difference in student performance in—guess what?—French classes.[4]

The evidence of harm is so massive that in 2014 the American Academy of Pediatrics (AAP) issued a policy statement calling for middle schools and high schools to begin no earlier than 8:30 a.m.[5] A few years later, the CDC added its voice, concluding that "delaying school start times has the potential for the greatest population impact" in boosting teenage learning and well-being.

Many school districts—from Dobbs Ferry, New York, to Houston, Texas, to Melbourne, Australia—have heeded the evidence and shown impressive results. For example, one study examined three years of data on 9,000 students from eight high schools in Minnesota, Colorado, and Wyoming that had changed their schedules to begin school after 8:35 a.m. At these schools, attendance rose and tardiness declined. Students earned higher grades "in core subject areas of math, English, science and social studies" and improved their performance on state and national standardized tests. At one school, the number of car crashes for teen drivers fell by 70 percent after it pushed its start time from 7:35 a.m. to 8:55 a.m.[6]

Another study of 30,000 students across seven states found that two years after implementing a later start time high school graduation rates increased by more than 11 percent.[7] One review of the start-time literature concludes that later start times correspond to "improved attendance, less tardiness . . . and better grades."[8] Equally important, students fare better not just in the classroom but also in many other domains of their lives. Considerable research finds that delaying school starting times improves motivation, boosts emotional well-being, reduces depression, and lessens impulsivity.[9]

The benefits aren't just for high school students; they extend to college students as well. At the United States Air Force Academy, delaying the school day's start time by fifty minutes improved academic performance; the later that first period began, the higher the students' grades.[10] In fact, a study of university students in both the United States and the United Kingdom, published in *Frontiers in Human Neuroscience*, concludes that the optimal time for most college classes is after 11 a.m.[11]

Even the price is right. When an economist studied the Wake County, North Carolina, school system, he found that "a 1 hour delay in start time increases standardized test scores on both math and reading tests by three percentile points," with the strongest effects on the weakest students.[12] But being an economist, he also calculated the cost-benefit ratio of changing the schedule and concluded that later start times delivered more bang for the educational buck than almost any other initiative available to policy makers, a view echoed by a Brookings Institution analysis.[13]

Yet the pleas of the nation's pediatricians and its top public-health officials, as well as the experiences of schools that have challenged the status quo, have been largely ignored. Today, fewer than one in five U.S. middle schools and high schools follow the AAP's recommendation to begin school after 8:30 a.m. The average start time for American adolescents remains 8:03 a.m., which means huge numbers of schools start in the 7 a.m. hour.[14]

Why the resistance? A key reason is that starting later is inconvenient for adults. Administrators must reconfigure bus schedules. Parents might not be able to drop off their kids on the way to work. Teachers must stay later in the afternoon. Coaches might have less time for sports practices.

But beneath those excuses is a deeper, and equally troubling, explanation. We simply don't take issues of *when* as seriously as we take questions of *what*. Imagine if schools suffered the same problems wrought by early start times—stunted learning and worsening health—but the cause was an airborne virus that was infecting classrooms. Parents would march to the schoolhouse to demand action and quarantine their children at home until the problem was solved. Every school district would snap into action. Now imagine if we could eradicate that virus and protect all those students with an already-known, reasonably priced, simply administered vaccine. The change would have already happened. Four out of five American school districts—more than 11,000—wouldn't be ignoring the evidence and manufacturing excuses. Doing so would be morally repellent and politically untenable. Parents, teachers, and entire communities wouldn't stand for it.

The school start time issue isn't new. But because it's a *when* problem rather than a *what* problem such as viruses or terrorism, too many people find it easy to dismiss. "What difference can one hour possibly make?" ask the forty- and fifty-year-olds. Well, for some students, it means the difference between dropping out and completing high school. For others, it's the difference between struggling with academics and mastering math and language courses—which can later affect their likelihood of going to college or finding a good job. In some cases, this small difference in timing could alleviate suffering and even save lives.

Starts matter. We can't always control them. But this is one area where we can and therefore we must.

STARTING AGAIN

At some point in your life, you probably made a New Year's resolution. On January 1 of some year, you resolved to drink less, exercise more, or call your mother every Sunday. Maybe you kept your resolution and rectified your health and family relations. Or maybe, by February, you were pasted on the couch watching *Legend of Kung Fu Rabbit* on Netflix while downing a third glass of wine and ducking Mom's Skype requests. Regardless of your resolution's fate, though, the date you chose to motivate yourself reveals another dimension of the power of beginnings.

The first day of the year is what social scientists call a "temporal landmark."[15] Just as human beings rely on landmarks to navigate space—"To get to my house, turn left at the Shell station"—we also use landmarks to navigate time. Certain dates function like that Shell station. They stand out from the ceaseless and forgettable march of other days, and their prominence helps us find our way.

In 2014 three scholars from the Wharton School of the University of Pennsylvania published a breakthrough paper in the science of timing that broadened our understanding of how temporal landmarks operate and how we can use them to construct better beginnings.

Hengchen Dai, Katherine Milkman, and Jason Riis began by analyzing eight and a half years of Google searches. They discovered that searches for the word "diet" always soared on January 1—by about 80 percent more than on a typical day. No surprise, perhaps. However, searches also spiked at the start of every calendar cycle—the first day of every month and the first day of every week. Searches even climbed 10 percent on the first day after a federal holiday. Something about days that represented "firsts" switched on people's motivation.

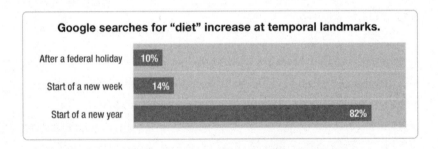

Google searches for "diet" increase at temporal landmarks.

After a federal holiday	10%
Start of a new week	14%
Start of a new year	82%

The researchers found a similar pattern at the gym. At a large northeastern university where students had to swipe a card to enter workout facilities, the researchers collected more than a year's worth of data on daily gym attendance. As with the Google searches, gym visits increased "at the start of each new week, month, and year." But those weren't the only dates that got students out of the dorm and onto a treadmill. Undergraduates "exercised more both at the start of a new semester . . . and on the first day after a school break." They also hit the gym more immediately after a birthday—with one glaring exception: "Students turning 21 tend to decrease their gym activity following their birthday."[16]

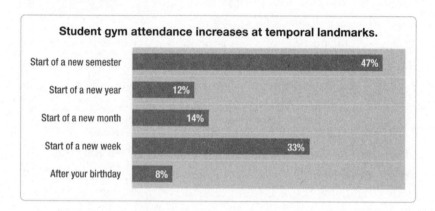

Student gym attendance increases at temporal landmarks.

Start of a new semester	47%
Start of a new year	12%
Start of a new month	14%
Start of a new week	33%
After your birthday	8%

For the Google searchers and college exercisers, some dates on the calendar were more significant than others. People were using them to "demarcate the passage of time," to end one period and begin another with a clean slate. Dai, Milkman, and Riis called this phenomenon the "fresh start effect."

To establish a fresh start, people used two types of temporal landmarks—social and personal. The social landmarks were those that everyone shared: Mondays, the beginning of a new month, national holidays. The personal ones were unique to the individual: birthdays, anniversaries, job changes. But whether social or personal, these time markers served two purposes.

First, they allowed people to open "new mental accounts" in the same way that a business closes the books at the end of one fiscal year and opens a fresh ledger for the new year. This new period offers a chance to start again by relegating our old selves to the past. It disconnects us from that past self's mistakes and imperfections, and leaves us confident about our new, superior selves. Fortified by that confidence, we "behave better than we have in the past and strive with enhanced fervor to achieve our aspirations."[17] In January advertisers often use the phrase "New Year, New You." When we apply temporal landmarks, that's what's going on in our heads.[18] *Old Me never flossed. But New Me, reborn on the first day back from summer vacation, will be a fiend about oral hygiene.*

The second purpose of these time markers is to shake us out of the tree so we can glimpse the forest. "Temporal landmarks interrupt attention to day-to-day minutiae, causing people to take a big picture view of their lives and thus focus on achieving their goals."[19] Think about those spatial landmarks again. You might drive for miles and barely notice your surroundings. But that glowing Shell station on the corner makes you pay attention. It's the same with fresh start dates. Daniel Kahneman draws a distinction between thinking fast (making decisions anchored in instinct and distorted by cognitive biases) and thinking slow (making decisions rooted in reason and guided by careful deliberation). Temporal landmarks slow our thinking, allowing us to deliberate at a higher level and make better decisions.[20]

The implications of the fresh start effect, like the forces that propel it, are also personal and social. Individuals who get off to a stumbling start—at a new job, on an important project, or in trying to improve their health—can alter their course by using a temporal landmark to start again. People can, as the Wharton researchers write, "strategically [create] turning points in their personal histories."[21]

Take Isabel Allende, the Chilean-American novelist. On January 8, 1981, she wrote a letter to her deathly ill grandfather. That letter formed the foundation of her first novel, *The House of the Spirits*. Since then, she has started each subsequent novel on that same date, using January 8 as a temporal landmark to make a fresh start on a new project.[22]

In later research, Dai, Milkman, and Riis found that imbuing an otherwise ordinary day with personal meaning generates the power to activate new beginnings.[23] For instance, when they framed March 20 as the first day of spring, the date offered a more effective fresh start than simply identifying it as the third Thursday in March. For Jewish participants in their study, reframing October 5 as the first day after Yom Kippur was more motivating than thinking of it as the 278th day of the year. Identifying one's own personally meaningful days—a child's birthday or the anniversary of your first date with your partner—can erase a false start and help us begin anew.

Organizations, too, can enlist this technique. Recent research has shown that the fresh start effect applies to teams.[24] Suppose a company's new quarter has a rough beginning. Rather than waiting until the next quarter, an obvious fresh start date, to smooth out the mess, leaders can find a meaningful moment occurring sooner—perhaps the anniversary of the launch of a key product—that would relegate previous screwups to the past and help the team get back on track. Or suppose some employees are not regularly contributing to their retirement accounts or failing to attend important training sessions. Sending them reminders on their birthdays rather than on some other day could prompt them to start acting. Consumers might also be more open to messages on days framed as fresh starts, Riis found.[25] If you're trying to encourage people to eat healthier, a campaign calling for Meatless Mondays will be far more effective than one advocating Vegan Thursdays.

New Year's Day has long held a special power over our behavior. We turn the page on the calendar, glimpse all those beautiful empty squares, and open a new account book on our lives. But we typically do that unwittingly, blind to the psychological mechanisms we're relying on. The fresh start effect allows us to use the same technique, but with awareness and intention, on multiple days. After all, New Year's resolutions are hardly foolproof. Research shows that a month into a new year only 64 percent of resolutions continue to be pursued.[26] Constructing our own temporal landmarks, especially those that are personally meaningful, gives us many more opportunities to recover from rough beginnings and start again.

STARTING TOGETHER

I n June of 1986, I graduated from college—unemployed. In July of 1986, I moved to Washington, D.C., to begin my postcollegiate life. By August of 1986, I'd found employment and was working in my first job. The elapsed time between receiving my diploma in a university auditorium and settling into my desk in downtown D.C. was less than sixty days. (And I didn't even spend all those days looking for work. Some of the time I was packing and moving. Some of it I was working at a bookstore to support myself during my brief job search.)

As much as I prefer to believe that my swift path from jobless graduate to youthful working stiff was due to my sterling credentials and winning personality, the more plausible reason is one that won't surprise you by now: timing. I graduated at an auspicious time. In 1986, the United States was surging out of a deep recession. The national unemployment rate that year was 7 percent—not an amazing figure but a huge drop from 1982 and 1983, when the jobless rate reached nearly 10 percent. This meant that it was simpler for me to find a job than for those who'd entered the job market just a few years earlier. It's not that complicated: You don't need a degree in economics to grasp that finding work is easier when the unemployment rate is 7 percent than when it's 10 percent. However, you have to be a pretty good economist to understand that the advantage I gained from the pure luck of beginning my work life in a relative boom lasted well beyond my first job.

Lisa Kahn is more than a pretty good economist. She made her mark in the economics world by studying people like me—white males who graduated from college in the 1980s. Kahn, who teaches at the Yale School of Management, harvested data from the National Longitudinal Survey of Youth, which each year asks a representative sample of American young people questions about their education, health, and employment. From the data, she selected white men who had graduated from college between 1979 and 1989—and examined what happened to them over the next twenty years.*

Her big discovery: *When* these men began their careers strongly determined where they went and how far they traveled. Those who entered the job market in weak economies earned less at the beginning of their careers than those who started in strong economies— no big surprise. But this early disadvantage didn't fade. It persisted for as long as twenty years.

"Graduating from college in a bad economy has a long-run, negative impact on wages," she writes. The unlucky graduates who'd begun their careers in a sluggish economy earned less straight out of school than the lucky ones like me who'd graduated in robust times—and it often took them *two decades* to catch up. On average, even after fifteen years of work, people who'd graduated in high unemployment years were still earning 2.5 percent less than those who'd graduated in low unemployment years. In some cases, the wage difference between graduating in an especially strong year versus an especially weak one was 20 percent—not just immediately after college but even when these men had reached their late thirties.[27] The total cost, in inflation-adjusted terms, of graduating in a bad year rather than a good year averaged about $100,000. Timing wasn't everything—but it was a six-figure thing.

---

* Kahn chose white males because their employment and earnings prospects are less affected by race and sex discrimination and because their career paths are less likely to be interrupted by having children. That allowed her to separate economic conditions from factors such as skin color, ethnicity, and gender.

Once again, beginnings set off a cascade that proved difficult to restrain. A large portion of one's lifetime wage growth occurs in the first ten years of a career. Starting with a higher salary puts people on a higher initial trajectory. But that's only the first advantage. The best way to earn more is to match your particular skills to an employer's particular needs. That rarely happens in one's first job. (My own first job, for instance, turned out to be a disaster.) So people quit jobs and take new ones—often every few years—to get the match right. Indeed, one of the fastest routes to higher pay early in a career is to switch jobs fairly often. However, if the economy is listless, changing jobs is difficult. Employers aren't hiring. And that means people who enter the labor market in a downturn are often stuck longer in jobs that aren't a good match for their skills. They can't switch employers easily, so it takes longer to locate a better match and begin the upward march to higher pay. What Kahn discovered in the job market is what chaos and complexity theorists have long known: In any dynamic system, the initial conditions have a huge influence over what happens to the inhabitants of that system.[28]

Other economists have likewise found that beginnings exert a powerful but invisible influence on people's livelihoods. In Canada, one study found that "the cost of recessions for new graduates is substantial and unequal." Unlucky graduates suffer "persistent earnings declines lasting ten years," with the least skilled workers suffering the most.[29] The cut may eventually heal, but it leaves a scar. A 2017 study found that economic conditions at the beginning of managers' careers have lasting effects on their becoming a CEO. Graduating in a recession makes it tougher to find a first job, which makes it more likely that aspiring managers will take a job at a smaller private firm than a large public company—which means they begin climbing a shorter ladder rather than a taller one. Those who began their careers during a recession do become CEOs—but they become CEOs of smaller firms and earn less money than their counterparts who graduated during boom years. Recession graduates, the research found, also have more conservative management styles, perhaps another legacy of less certain beginnings.[30]

Research on Stanford MBAs has found that the state of the stock market at the time of graduation shapes these graduates' lifetime earnings. The chain of logic and circumstance here has three links. First, students are more likely to take jobs on Wall Street when they graduate in a bull market. By contrast, in bear markets, a sizable portion of graduates choose alternatives—consulting, entrepreneurship, or working for nonprofits. Second, people who work on Wall Street tend to remain working on Wall Street. Third, investment bankers and other financial professionals generally outearn those in other fields. As a result, "a person who graduates in a bull market" and goes into investment banking earns an additional $1.5 to $5 million more than "that same person would have earned if he or she had graduated during a bear market" and therefore had shied away from a Wall Street job.[31]

My sleep will remain undisturbed knowing that a swerving stock market steered some elite MBAs to jobs at McKinsey or Bain rather than at Goldman Sachs or Morgan Stanley and thereby left them extremely rich rather than insanely wealthy. But the effects of beginnings on a large swath of the workforce is more troubling, especially since the early data on those who entered the job market during the 2007–2010 Great Recession look especially dim. Kahn and two Yale colleagues have found that the negative impact on students who graduated during 2010 and 2011 "was double what we would have expected given past patterns."[32] The Federal Reserve Bank of New York, looking at these early indicators, warned that "those who begin their careers during such a weak labor market recovery may see *permanent* negative effects on their wages."[33]

This is a tough problem. If what you earn today depends heavily on the unemployment rate when you started working rather than on the unemployment rate now, the previous two strategies in this chapter—starting right and starting again—are insufficient.[34] We can't solve the problem unilaterally, as with school starting times, and simply dictate that everyone will begin her career in a healthy economy. Nor can we solve it individually by exhorting people to recover from their slow start by looking for a new job on the day after their birthday. On this sort of problem, we must start together. And two previous smart solutions offer some guidance.

For many years, teaching hospitals in the United States confronted what was known as the "July effect." Each July, a fresh group of medical school graduates began their careers as physicians. Although these men and women had little experience beyond the class-room, teaching hospitals often gave them considerable responsibility for treating patients. That was how they learned their craft. The only downside of this approach is that patients often suffered from this on-the-job training—and July was the cruelest month. (In the UK, the month is later and the language more vivid. British physicians call the period when new doctors begin their jobs the "August kill-ing season.") For example, one study of more than twenty-five years of U.S. death certificates found that "in counties containing teaching hospitals, fatal medication errors spiked by 10% in July and in no other month. In contrast, there was no July spike in counties without teaching hospitals."[35] Other research in teaching hospitals found that patients in July and August had an 18 percent greater chance of surgery problems and a 41 percent greater chance of dying in surgery than patients did in April and May.[36]

However, in the last decade, teaching hospitals have worked to correct this. Instead of declaring bad beginnings an inevitable prob-lem for an individual, they made it a preventable problem for a group. Now, at teaching hospitals like the one I visited at the Uni-versity of Michigan, new residents begin their tenure by working as part of a team that includes seasoned nurses, physicians, and other professionals. By starting together, hospitals like this one have dramatically reduced the July effect.

Or consider babies born to young mothers in low-income neighborhoods. Children in those circumstances often suffer terrible beginnings. But one effective solution has been to ensure that mother and baby don't start alone. A national program called Nurse-Family Partnership, launched in the 1970s, sends nurses to visit mothers and help them get their babies off to a better beginning. The program, now in eight hundred U.S. municipalities, has also subjected itself to rigorous outside evaluation—with promising results. Nurse visits reduce infant mortality rates, limit behavior and attention problems, and minimize families' reliance on food stamps and other social welfare programs.[37] They've also boosted children's health and learning, improved breast-feeding and vaccination rates, and increased the chances mothers will seek and keep paid work.[38] Many European nations provide such visits as a matter of policy. Whether the reasons are moral (these programs save lives) or financial (these programs save money over the long term), the principle remains the same: Instead of forcing vulnerable people to fend for themselves, everyone does better by starting together.

We can apply similar principles to the problem that some people, through no fault of their own, begin their careers in lousy economies. We can't dismiss this issue: "Oh, that's just bad timing. Nothing we can do about that." Instead, we should recognize that having a lot of people earning too little or struggling to make their way affects all of us—in the form of fewer customers for what we're selling and higher taxes to deal with the consequences of limited opportunities. One solution might be for governments and universities to institute a student-loan-forgiveness program keyed to the unemployment rate. If the unemployment rate topped, say, 7.5 percent, some portion of a newly graduating student's loan would be forgiven. Or perhaps if the unemployment rate ticked above a certain mark, university or federal funds would be unlocked to pay for career counselors to help new graduates trek their way across the newly rocky terrain—in much the same way the federal government deploys sandbags and the Army Corps of Engineers to regions beset by floods.

The goal here is to recognize that slow-moving *when* problems have all the gravity of fast-moving *what* calamities—and deserve the same collective response.

M ost of us have harbored a sense that beginnings are significant. Now the science of timing has shown that they're even more powerful than we suspected. Beginnings stay with us far longer than we know; their effects linger to the end.

That's why, when we tackle challenges in our lives—whether losing a few pounds or helping our kids learn or ensuring that our fellow citizens aren't caught in the downdraft of circumstance—we need to expand our repertoire of responses and include *when* alongside *what*. Armed with the science, we can do a much better job of starting right—in schools and beyond. Knowing how our minds reckon with time can help us use temporal landmarks to recover from false starts and make fresh ones. And understanding how unfair—and enduring—rough beginnings can be might stir us to start together more often.

Shifting our focus—and giving *when* the same weight as *what*— won't cure all our ills. But it's a good beginning.

Time Hacker's Handbook

• CHAPTER 3 •

AVOID A FALSE START WITH A PREMORTEM

The best way to recover from a false start is to avoid one in the first place. And the best technique for doing that is something called a "premortem."

You've probably heard of a postmortem—when coroners and physicians examine a dead body to determine the cause of death. A premortem, the brainchild of psychologist Gary Klein, applies the same principle but shifts the exam from after to before.[1]

Suppose you and your team are about to embark on a project. Before the project begins, convene for a premortem. "Assume it's eighteen months from now and our project is a complete disaster," you say to your team. "What went wrong?" The team, using the power of prospective hindsight, offers some answers. Maybe the task wasn't clearly defined. Maybe you had too few people, too many people, or the wrong people. Maybe you didn't have a clear leader or realistic objectives. By imagining failure in advance—by thinking through what might cause a false start—you can anticipate some of the potential problems and avoid them once the actual project begins.

As it happens, I conducted a premortem before I began this book.

I projected two years from the start date and imagined that I'd written a terrible book or, worse, hadn't managed to write a book at all. Where did I go awry? After looking at my answers, I realized I had to be vigilant about writing every day, saying no to every outside obligation so I didn't get distracted, keeping my editor informed of my progress (or lack thereof), and enlisting his help early in untangling any conceptual knots. Then I wrote down the positive versions of these insights—for example, "I worked on the book all morning every morning at least six days a week with no distractions and no exceptions"—on a card that I posted near my desk.

The technique allowed me to make mistakes in advance in my head rather than in real life on a real project. Whether this particular premortem was effective I'll leave to you, dear reader. But I encourage you to try it to avoid your own false starts.

EIGHTY-SIX DAYS IN THE YEAR WHEN YOU CAN MAKE A FRESH START

You've read about temporal landmarks and how we can use them to fashion fresh starts. To help you on that quest for an ideal day to begin that novel or commence training for a marathon, here are eighty-six days that are especially effective for making a fresh start:

- The first day of the month (twelve)
- Mondays (fifty-two)
- The first day of spring, summer, fall, and winter (four)
- Your country's Independence Day or the equivalent (one)
- The day of an important religious holiday—for example, Easter, Rosh Hashanah, Eid al-Fitr (one)
- Your birthday (one)

- A loved one's birthday (one)
- The first day of school or the first day of a semester (two)
- The first day of a new job (one)
- The day after graduation (one)
- The first day back from vacation (two)
- The anniversary of your wedding, first date, or divorce (three)
- The anniversary of the day you started your job, the day you became a citizen, the day you adopted your dog or cat, the day you graduated from school or university (four)
- The day you finish this book (one)

WHEN SHOULD YOU GO FIRST?

Life isn't always a competition, but it is sometimes a *serial* competition. Whether you're one of several people interviewing for a job, part of a lineup of companies pitching for new business, or a contestant on a nationally televised singing program, *when* you compete can be just as important as what you do.

Here, based on several studies, is a playbook for when to go first—and when not to:

Four Situations When You *Should* Go First

1. If you're on a ballot (county commissioner, prom queen, the Oscars), being listed first gives you an edge. Researchers have studied this effect in thousands of elections—from school board to city council, from California to Texas—and voters consistently preferred the first name on the ballot.[2]
2. If you're *not* the default choice—for example, if you're pitching against a firm that already has the account you're seeking—going first can help you get a fresh look from the decision-makers.[3]

3. If there are relatively few competitors (say, five or fewer), going first can help you take advantage of the "primacy effect," the tendency people have to remember the first thing in a series better than those that come later.[4]

4. If you're interviewing for a job and you're up against several strong candidates, you might gain an edge from being first. Uri Simonsohn and Francesca Gino examined more than 9,000 MBA admissions interviews and found that interviewers often engage in "narrow bracketing"—assuming small sets of candidates represent the entire field. So if they encounter several strong applicants early in the process, they might more aggressively look for flaws in the later ones.[5]

Four Situations When You *Should Not* Go First

1. If you *are* the default choice, *don't* go first. Recall from the previous chapter: Judges are more likely to stick with the default late in the day (when they're fatigued) rather than early or after a break (when they're revived).[6]

2. If there are many competitors (not necessarily strong ones, just a large number of them), going later can confer a small advantage and going last can confer a huge one. In a study of more than 1,500 live *Idol* performances in eight countries, researchers found that the singer who performed last advanced to the next round roughly 90 percent of the time. An almost identical pattern occurs in elite figure skating and even in wine tastings. At the beginning of competitions, judges hold an idealized standard of excellence, say social psychologists Adam Galinsky and Maurice Schweitzer. As the competition proceeds, a new, more realistic baseline develops, which favors later competitors, who gain the added advantage of seeing what others have done.[7]

3. If you're operating in an uncertain environment, *not* being first can work to your benefit. If you don't know what the decision-maker

expects, letting others proceed could allow the criteria to sharpen into focus both for the selector and you.[8]

4. If the competition is meager, going toward the end can give you an edge by highlighting your differences. "If it was a weak day with many bad candidates, it's a really good idea to go last," says Simonsohn.[9]

FOUR TIPS FOR MAKING A FAST START IN A NEW JOB

You've read about the perils of graduating in a recession. We can't do much to avoid that fate. But whenever we begin a new job—in a recession or a boom—we can influence how much we enjoy the job and how well we do. With that in mind, here are four research-backed recommendations for how to make a fast start in a new job.

1. Begin before you begin.

Executive advisor Michael Watkins recommends picking a specific day and time when you visualize yourself "transforming" into your new role.[10] It's hard to get a fast start when your self-image is stuck in the past. By mentally picturing yourself "becoming" a new person even before you enter the front door, you'll hit the carpet running. This is especially true when it comes to leadership roles. According to former Harvard professor Ram Charan, one of the toughest transitions lies in going from a specialist to a generalist.[11] So as you think about your new role, don't forget to see how it connects to the bigger picture. For one of the ultimate new jobs— becoming president of the United States—research has shown that one of the best predictors of presidential success is how early the transition began and how effectively it was handled.[12]

2. Let your results do the talking.

A new job can be daunting because it requires establishing yourself in the organization's hierarchy. Many individuals overcompensate for their initial nervousness and assert themselves too quickly and too soon. That can be counterproductive. Research from UCLA's Corinne Bendersky suggests that over time extroverts lose status in groups.[13] So, at the outset, concentrate on accomplishing a few meaningful achievements, and once you've gained status by demonstrating excellence, feel free to be more assertive.

3. Stockpile your motivation.

On your first day in a new role, you'll be filled with energy. By day thirty? Maybe less so. Motivation comes in spurts—which is why Stanford psychologist B. J. Fogg recommends taking advantage of "motivation waves" so you can weather "motivation troughs."[14] If you're a new salesman, use motivation waves to set up leads, organize calls, and master new techniques. During troughs, you'll have the luxury of working at your core role without worrying about less interesting peripheral tasks.

4. Sustain your morale with small wins.

Taking a new job isn't exactly like recovering from an addiction, but programs such as Alcoholics Anonymous do offer some guidance. They don't order members to embrace sobriety forever but instead ask them to succeed "24 hours at a time," something Karl Weick noted in his seminal work on "small wins."[15] Harvard professor Teresa Amabile concurs. After examining 12,000 daily diary entries by several hundred workers, she found that the single largest motivator was making progress in meaningful work.[16] Wins needn't be large to be meaningful. When you enter a new role, set up small "high-probability" targets and celebrate when you hit them. They'll give you the motivation and energy to take on more daunting challenges further down the highway.

WHEN SHOULD YOU GET MARRIED?

One of the most important beginnings many of us make in life is getting married. I'll leave it to others to recommend whom you should marry. But I can give you some guidance about when to tie the knot. The science of timing doesn't provide definitive answers, but it does offer three general guidelines:

1. Wait until you're old enough (but not too old).

It's probably no surprise that people who marry when they're very young are more likely to divorce. For instance, an American who weds at twenty-five is 11 percent less likely to divorce than one who marries at age twenty-four, according to an analysis by University of Utah sociologist Nicholas Wolfinger. But waiting too long has a downside. Past the age of about thirty-two—even after controlling for religion, education, geographic location, and other factors—the odds of divorce *increase* by 5 percent per year for at least the next decade.[17]

2. Wait until you've completed your education.

Couples tend to be more satisfied with their marriages, and less likely to divorce, if they have more education before the wedding. Consider two couples. They're the same age and race, have comparable incomes, and have attended the same total amount of school. Even among these similar couples, the pair who weds after completing school is more likely to stay together.[18] So finish as much education as you can before getting hitched.

3. Wait until your relationship matures.

Andrew Francis-Tan and Hugo Mialon at Emory University found that couples that dated for at least one year before marriage were 20 percent less likely to divorce than those who made the move more quickly.[19] Couples that had dated for more than three years were even less likely to split up once they exchanged vows. (Francis-Tan and Mialon also found that the more a couple spent on its wedding and any engagement ring, the more likely they were to divorce.)

In short, for one of life's ultimate *when* questions, forget the romantics and listen to the scientists. Prudence beats passion.

4.

MIDPOINTS

What Hanukkah Candles and
Midlife Malaise Can Teach
Us About Motivation

When you are in the middle of a story it isn't a story
at all, but only a confusion; a dark roaring, a blindness,
a wreckage of shattered glass and splintered wood.

—MARGARET ATWOOD, *Alias Grace*

Our lives rarely follow a clear, linear path. More often, they're a series of episodes—with beginnings, middles, and ends. We often remember beginnings. (Can you picture your first date with your spouse or partner?) Endings also stand out. (Where were you when you heard that a parent, grandparent, or loved one had died?) But middles are muddy. They recede rather than reverberate. They get lost, well, in the middle.

Yet the science of timing is revealing that midpoints have powerful, though peculiar, effects on what we do and how we do it. Sometimes hitting the midpoint—of a project, a semester, a life— numbs our interest and stalls our progress. Other times, middles stir and stimulate; reaching the midpoint awakens our motivation and propels us onto a more promising path.

I call these two effects the "slump" and the "spark."

Midpoints can bring us down. That's the slump. But they can also fire us up. That's the spark. How can we identify the difference? And how, if at all, can we turn a slump into a spark? Finding the answers requires lighting some holiday candles, making a radio commercial, and revisiting one of college basketball's greatest games. But let's launch our inquiry with what many consider the ultimate physical, emotional, and existential midpoint droop: middle age.

THAT'S WHAT I LIKE ABOUT U

In 1965, an obscure Canadian psychoanalyst named Elliott Jaques published a paper in an equally obscure publication called the *International Journal of Psychoanalysis*. Jaques had been examining the biographies of prominent artists, including Mozart, Raphael, Dante, and Gauguin, and he noticed that an unusual number of them seemed to have died at age thirty-seven. Atop that flimsy factual foundation, he added a few floors of Freudian jargon, plopped a staircase of hazy clinical anecdotes in the center, and emerged with a fully constructed theory.

"In the course of the development of the individual," Jaques wrote, "there are critical phases which have the character of change points, or periods of rapid transition." And the least familiar but most crucial of these phases, he said, occurs around age thirty-five— "which I shall term the mid-life crisis."[1]

Kaboom!

The idea detonated. The phrase "midlife crisis" leaped onto magazine covers. It crept into TV dialogue. It launched dozens of Hollywood films and sustained the panel-discussion industry for at least two decades.[2]

"The central and crucial feature of the mid-life phase," Jaques said, was the "inevitability of one's own eventual personal death." When people reach the middle of their lives, they suddenly spy the Grim Reaper in the distance, which uncorks "a period of psychological disturbance and depressive breakdown."[3] Haunted by the specter of death, middle-aged people either succumb to its inevitability or radically redirect their course to avoid reckoning with it. The phrase infiltrated the global conversation with astonishing speed.

It remains part of the parlance today; the tableau of cultural clichés is as vivid as ever. We know what a midlife crisis looks like even when it's updated for contemporary times. Mom impulsively buys a cherry Maserati—in midlife crises, the cars are always red and sporty—and zooms away with her twenty-five-year-old assistant. Dad disappears with the pool boy to open a vegan café in Palau. A full half century after Jaques lobbed his conceptual grenade, the midlife crisis is everywhere.

Everywhere, that is, except in the evidence.

When developmental psychologists have looked for it in the laboratory or the field, they've largely come up empty. When pollsters have listened for it in public-opinion surveys, this supposed cri de coeur barely registers. Instead, during the last ten years, researchers have detected a quieter midlife pattern, one that arrives with remarkable consistency across the world and that reflects a broader truth about midpoints of every kind.

For example, in 2010 four social scientists, including Nobel Prize–winning economist Angus Deaton, took what they called "a snapshot of the age distribution of well-being in the United States." The team asked 340,000 interviewees to imagine themselves on a ladder with steps numbered from 0 at the bottom to 10 at the top. If the top

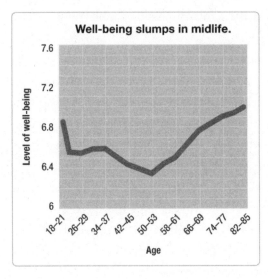

Well-being slumps in midlife.

step represented their best possible life, and the bottom the worst possible one, what step were they standing on now? (The question was a more artful way of asking, "On a scale of 0 to 10, how happy are you?") The results, even controlling for income and demographics, were shaped like a shallow U, as you can see in the chart. People in their twenties and thirties were reasonably happy, people in their forties and early fifties less so, and people from about fifty-five onward happier once again.[4]

Well-being in midlife didn't collapse in a cataclysmic, life-altering way. It just sagged.

This U-curve of happiness—a mild slump rather than a raging crisis—is an extremely robust finding. A slightly earlier study of more than 500,000 Americans and Europeans by economists David Blanchflower and Andrew Oswald found that well-being consistently slid around the middle of life. "The regularity is intriguing," they observe. "The U-shape is similar for males and females, and for each side of the Atlantic Ocean." But it wasn't merely an Anglo-American phenomenon. Blanchflower and Oswald also analyzed data from around the world and discovered something remarkable. "In total, we document a statistically significant U-shape in happiness or life satisfaction for 72 countries," they write, from Albania and Argentina through the nation-state alphabet to Uzbekistan and Zimbabwe.[5]

Study after study across an astonishing range of socioeconomic, demographic, and life circumstances has reached the same conclusion: Happiness climbs high early in adulthood but begins to slide downward in the late thirties and early forties, dipping to a low in the fifties.[6] (Blanchflower and Oswald found that "subjective well-being among American males bottoms out at an estimated 52.9 years."[7]) But we recover quickly from this slump, and well-being later in life often exceeds that of our younger years. Elliott Jaques was on the right track but aboard the wrong train. Something does indeed happen to us at midlife, but the actual evidence is far less dramatic than his original speculation.

But why? Why does this midpoint deflate us? One possibility is the disappointment of unrealized expectations. In our naïve twenties and thirties, our hopes are high, our scenarios rosy. Then reality trickles in like a slow leak in the roof. Only one person gets to be CEO—and it's not going to be you. Some marriages crumble—and yours, sadly, is one of them. That dream of owning a Premier League team becomes remote when you can barely cover your mortgage. Yet we don't remain in the emotional basement for long, because over time we adjust our aspirations and later realize that life is pretty good. In short, we dip in the middle because we're lousy forecasters. In youth, our expectations are too high. In older age, they're too low.[8]

However, another explanation is also plausible. In 2012, five scientists asked zookeepers and animal researchers in three countries to help them better understand the more than 500 great apes under their collective care. These primates—chimpanzees and orangutans—ranged from infants to older adults. The researchers wanted to know how they were doing. So they asked the human personnel to rate the apes' mood and well-being. (Don't laugh. The researchers explain that the questionnaire they used "is a well-established method for assessing positive affect in captive primates.") Then they matched those happiness ratings to the ages of the great

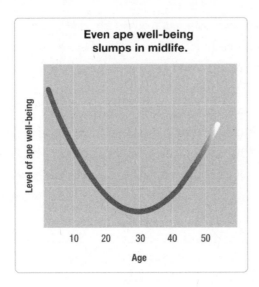

Even ape well-being slumps in midlife.

Level of ape well-being

Age

10   20   30   40   50

apes. The resulting chart is shown here.[9]

That raises an intriguing possibility: Could the midpoint slump be more biology than sociology, less a malleable reaction to circumstance than an immutable force of nature?

LIGHTING CANDLES AND CUTTING CORNERS

A traditional box of Hanukkah candles contains forty-four candles, a number determined with Talmudic precision. Hanukkah lasts eight consecutive nights, and Jews who celebrate the holiday mark their observance each evening by lighting candles positioned in a candleholder known as a menorah. On the first night, celebrants light one candle, two candles on the second night, and so on. Because observers light each candle with a helper candle, they end up using two candles on the first night, three on the second night, and eventually nine candles on the eighth night, yielding the following formula:

$$2 + 3 + 4 + 5 + 6 + 7 + 8 + 9 = 44$$

Forty-four candles means that when the holiday ends, the box will be empty. Yet, in Jewish households across the world, families routinely finish Hanukkah with candles left in the box.

What gives? How to solve this mystery of the lights?

Diane Mehta offers part of the answer. Mehta is a novelist and poet who lives in New York. Her mother is a Jew from Brooklyn, her father a Jain from India. She grew up in New Jersey, where she celebrated Hanukkah, eagerly lit the candles, and "got things like socks as gifts." When she had a son, he, too, loved lighting the candles. But as time passed—job changes, a divorce, the usual ups and downs of life—her candle lighting became less regular. "I start off getting excited," she told me. "But after a couple of days, I taper off." She doesn't light the candles when her son is staying with his dad rather than with her. But sometimes, toward the end of the holiday, she says, "I'll notice that it's still Hanukkah and will light the candles again. I'll say to my son, 'It's the last night. We should do it.'"

Mehta often begins Hanukkah with zest and ends with resolve but slacks in the middle. She sometimes neglects lighting candles on nights three, four, five, and six—and thus ends the holiday with candles still in the box. And she's not alone.

Maferima Touré-Tillery and Ayelet Fishbach are two social scientists who study how people pursue goals and adhere to personal standards. A few years ago, they were searching for a real-world domain in which to explore these two ideas when they realized that Hanukkah represented an ideal field study. They tracked the behavior of more than two hundred Jewish participants who observed the holiday, measuring whether—and, crucially, when—they lit the candles. After eight nights of collecting data, here's what they found:

On the first night, 76 percent of the participants lit the candles.

On the second night, the percentage dropped to 55.

On the ensuing nights, fewer than half the participants lit the candles—with the number climbing above 50 percent again only on night eight.

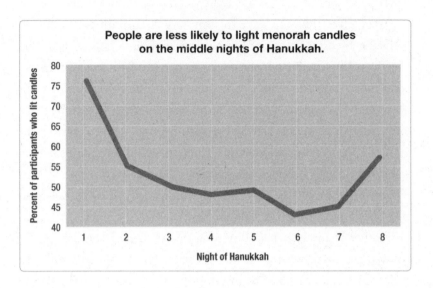

People are less likely to light menorah candles on the middle nights of Hanukkah.

Over the course of Hanukkah, the researchers conclude, "adherence to standards followed a U-shaped pattern."[10]

But perhaps this slump had an easy explanation. Maybe the less religious participants, unlike their more observant counterparts, were opting out in the middle and lowering the average. Touré-Tillery and Fishbach tested for that possibility. They found that the U-shaped pattern became more pronounced for the most religious participants. They were even more likely than others to light the candles on nights 1 and 8. But in the middle of Hanukkah, "their behavior was almost undistinguishable from that of less religious participants."[11]

The researchers surmised that what was going on was "signaling." We all want others to think well of us. And for some people, the lighting of Hanukkah candles, often done in front of others, is a signal of religious virtue. However, the celebrants believed the signals that mattered most, the ones that projected their images most powerfully, were those at the beginning and end. The middle didn't matter as much. And they turned out to be right. When Touré-Tillery and Fishbach conducted a subsequent experiment in which they asked people to assess the religiousness of three fictitious characters based on when those characters lit candles, "participants thought the persons who did not light the Menorah on the first and last night were less religious than the person who skipped the ritual on the fifth night."

In the middle, we relax our standards, perhaps because others relax their assessments of us. At midpoints, for reasons that are elusive but enlightening, we cut corners—as one last experiment shows. Touré-Tillery and Fishbach also engaged other participants in what they claimed was a test of how young adults perform on skills they hadn't used much since childhood. They handed people a stack of five cards, each of which had a shape drawn on it. The shape was always the same, but it was rotated into a different position on each card. They gave people scissors and asked them to cut out the shapes as carefully as possible. Then the researchers presented the cutout shapes to lab workers not involved in the experiment and asked them to rate, on a 1-to-10 scale, the cutting accuracy of the five shapes.

The result? Participants' scissor skills rose at the beginning and

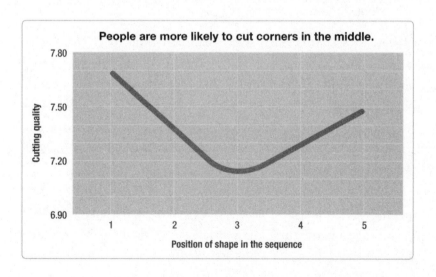

123

end but slumped in the middle. "In the domain of performance standards, we thus found that participants were more likely to literally cut corners in the middle of the sequence rather than at the beginning and end."

Something takes over in the middle—something that seems more like a celestial power than an individual choice. Just as the bell curve represents one natural order, the U-curve represents another. We can't eliminate it. But as with any force of nature—thunderstorms, gravity, the human drive to consume calories—we can mitigate some of its harms. The first step is simply awareness. If the midlife droop is inevitable, just knowing that eases some of the pain, as does knowing that the state is not permanent. If we're aware that our standards are likely to sink at the midpoint, that knowledge can help us temper the consequences. Even if we can't hold off biology and nature, we can prepare for their ramifications.

But we also have another option. We can use a little biology to fight back.

THE UH-OH EFFECT

The best scientists often start small and think big. That's what Niles Eldredge and Stephen Jay Gould did. In the early 1970s, both were young paleontologists. Eldredge studied a breed of trilobite that lived more than 300 million years ago. Gould, meanwhile, concentrated his efforts on two varieties of Caribbean land snails. But when Eldredge and Gould collaborated, as they did in 1972, their puny subjects led them to a gargantuan insight.

At the time, most biologists believed in a theory called "phyletic gradualism," which held that species evolve slowly and incrementally. Evolution, the thinking went, moves gradually—over millions upon millions of years—Mother Nature working steadily with

Father Time. Eldredge and Gould, however, saw something different in the fossil record of the arthropods and mollusks they were studying. The evolution of species sometimes advanced as sluggishly as the snails themselves. But at other moments, it exploded. Species experienced long periods of stasis that were interrupted by sudden bursts of change. Afterward, the newly transformed species remained stable for another long stretch—until another eruption abruptly altered its course once again. Eldredge and Gould called their new theory "punctuated equilibrium."[12] Evolution's path wasn't a smooth upward climb. The true trajectory was less linear: periods of dull stability punctuated by swift explosions of change. The Eldredge-Gould theory was itself a form of punctuated equilibrium—a massive conceptual explosion that interrupted a previously sleepy stretch in evolutionary biology and redirected the field down an alternative path.

A decade later, a scholar named Connie Gersick was beginning to study another organism (human beings) in its natural habitat (conference rooms). She tracked small groups of people working on projects—a task force at a bank developing a new type of account, hospital administrators planning a one-day retreat, university faculty and administrators designing a new institute for computer science—from their very first meeting to the moment they reached their final deadline. Management thinkers believed that teams working on projects moved gradually through a series of stages—and Gersick believed that by videotaping all the meetings and transcribing every word people uttered she could understand these consistent team processes in a more granular way.

What she found instead was inconsistency. Teams did not progress steadily through a universal set of stages. They used wildly diverse and idiosyncratic approaches to getting work done. The hospital team evolved differently from the banking team, which evolved differently from the computer science team. However, she wrote, what remained the same, even when everything else was diverging, was "the *timing* of when groups formed, maintained, and changed."[13]

Each group first went through a phase of prolonged inertia. The teammates got to know one another, but they didn't accomplish much. They talked about ideas but didn't move forward. The clock ticked. The days passed.

Then came a sudden transition. "In a concentrated burst of changes, groups dropped old patterns, reengaged with outside supervisors, adopted new perspectives on their work, and made dramatic progress," Gersick found. After the initial inert phase, they entered a new heads-down, locked-in phase that executed the plan and hurtled toward the deadline. But even more interesting than the burst itself was when it arrived. No matter how much time the various teams were allotted, "each group experienced its transition at the same point in its calendar—precisely halfway between its first meeting and its official deadline."

The bankers made their leap forward in designing a new account on "the 17th day of a 34-day span." The hospital administrators took off in a new, more productive direction in week six of a twelve-week assignment. So it went for every team. "As each group approached the midpoint between the time it started work and its deadline, it underwent great change," Gersick wrote. Groups didn't march toward their goals at a steady, even pace. Instead, they spent considerable time accomplishing almost nothing—until they experienced a surge of activity that always came at "the temporal midpoint" of a project.[14]

Since Gersick obtained results she didn't expect, and since those results ran counter to the prevailing view, she searched for a way to understand them. "The paradigm through which I came to interpret the findings resembles a relatively new concept from the field of natural history that has not heretofore been applied to groups: punctuated equilibrium," she wrote. Like those trilobites and snails, teams of human beings working together didn't progress gradually. They experienced extended periods of inertia—interrupted by swift bursts of activity. But in the case of humans, whose time horizons spanned a few months of work rather than millions of years of evolution, equilibrium always had the same punctuation mark: a midpoint.

For example, Gersick studied one group of business students given eleven days to analyze a case and prepare an explanatory paper. The teammates dickered and bickered at first and resisted outside advice. But on day six of their work—the precise midpoint of their project—the issue of timing parachuted into the conversation. "We're very short on time," warned one member. Shortly after that comment, the group abandoned its unpromising initial approach and generated a revised strategy that it pursued to the end. At the halfway mark in this team and the others, Gersick wrote, members felt "a new sense of urgency."

Call it the "uh-oh effect."

When we reach a midpoint, sometimes we slump, but other times we jump. A mental siren alerts us that we've squandered half of our time. That injects a healthy dose of stress—*Uh-oh, we're running out of time!*—that revives our motivation and reshapes our strategy.

In subsequent research, Gersick confirmed the power of the uh-oh effect. In one experiment, she assembled eight teams of MBA students and assigned them, after fifteen or twenty minutes of reading a design brief, to create a radio commercial in one hour. Then, as in her earlier work, she videotaped the interactions and transcribed the conversations. Every group made an uh-oh comment ("Okay, now we've reached the halfway point. Now we're *really* in trouble.") between twenty-eight and thirty-one minutes through the one-hour project. And six of these eight teams made their "most significant progress" during a "concentrated midpoint burst."[15]

She found the same dynamic over longer periods. In other research, she spent a year following a venture-capital-backed start-up company that she called M-Tech. Entire companies don't have the finite lives or specific deadlines of small project teams. Yet she found that M-Tech "showed many of the same basic temporally regulated punctuational patterns as project groups show, on a more sophisticated, deliberate level." That is, M-Tech's CEO scheduled all the company's key planning and evaluation meetings in July, the midpoint of the calendar year, and used what he learned to redirect M-Tech's second-half strategy.

"Midyear transitions, like midpoint transitions in groups, significantly shaped M-Tech's history," Gersick wrote. These breaks in time interrupted ongoing tactics and strategies and provided opportunities for management to evaluate and alter the company's course."[16]

Midpoints, as we're seeing, can have a dual effect. In some cases, they dissipate our motivation; in other cases, they activate it. Sometimes they elicit an "oh, no" and we retreat; other times, they trigger an "uh-oh" and we advance. Under certain conditions, they bring the slump; under others, they deliver the spark.

Think of midpoints as a psychological alarm clock. They're effective only when we set the alarm, when we can hear its annoying bleep, bleep, bleep go off, and when we don't hit the snooze button. But with midpoints, as with alarm clocks, the most motivating wake-up call is one that comes when you're running slightly behind.

HALFTIME SHOW

In the fall of 1981, a nineteen-year-old freshman from Kingston, Jamaica, by way of Cambridge, Massachusetts, walked onto the campus of Georgetown University in Washington, D.C. Patrick

Ewing didn't look like most first-year students. He was tall. Toweringly, staggeringly, monumentally tall. Yet he was also graceful, a young man who moved with the fluid quickness of a sprinter.

Ewing had come to Georgetown to help Coach John Thompson establish the school as a national basketball power. And from day one, Ewing was a transforming presence on the court. "A moving giant," the *New York Times* called him. "A center for the ages," said another newspaper. "A 7-foot monster child" who could devour opponent offenses like a "human PAC-MAN," *Sports Illustrated* gushed.[17] Ewing quickly made Georgetown one of the nation's top defensive teams. During his freshman season, the Hoyas won thirty games, a school record. For the first time in thirty-nine years, they reached the National Collegiate Athletic Association Final Four, where they won their semifinal game and found themselves playing for the national championship.*

Georgetown's opponent in that 1982 NCAA championship game was the University of North Carolina Tar Heels, led by All-American forward James Worthy and coached by Dean Smith. Dean Smith was a well-regarded coach but also a snakebit one. He had coached the Tar Heels for twenty-one years, taken them to the Final Four six times, and advanced to three finals. But to the dismay of his basketball-crazed state, he'd never brought home a national title. In tournament games, opposing fans had taken to heckling him with cries of "Choke, Dean, choke."

On the last Monday night of March, Smith's Tar Heels and Thompson's Hoyas faced off in the Louisiana Superdome in front of more than 61,000 fans, "the largest crowd ever to see a game in the Western Hemisphere."[18] Ewing intimidated from the outset, although not always in a productive way. North Carolina's first four

---

* During Ewing's four seasons at Georgetown, the Hoyas made the NCAA finals three times.

scores came on goaltending calls against Ewing. (Ewing illegally interfered with the ball as it was heading into the basket, something only a player of his size typically can do.) North Carolina didn't actually put the ball into the hoop for the first eight minutes of the game.[19] Ewing blocked shots, sunk free throws, and would eventually score twenty-three points. But North Carolina kept it close. With forty seconds left in the first half, Ewing raced eighty feet down the court on a fast break and slammed a dunk so thunderous that it nearly buckled the floorboards. At halftime, Georgetown led 32 to 31, a good omen. In the previous forty-three NCAA finals, the team ahead at the half had won thirty-four of them, an 80 percent success rate. During its regular season, Georgetown had a 26–1 record in games where it held a halftime lead.

Halftimes in sports represent another kind of midpoint—a specific moment in time when activity stops and teams formally reassess and recalibrate. But sports halftimes differ from life, or even project, midpoints on one important dimension: At this midpoint, the trailing team confronts harsh mathematical reality. The other team has more points. That means only matching them in the second half will guarantee a loss. The team that's behind must now not only outscore its opponent, it must also outscore the opposition by more than the amount it's trailing. A team ahead at halftime—in any sport—is more likely than its opponent to win the game. This has little to do with the limits of personal motivation and everything to do with the heartlessness of probability.

However, there's an exception—one peculiar circumstance where motivation seems to trump mathematics.

Jonah Berger of the University of Pennsylvania and Devin Pope of the University of Chicago analyzed more than 18,000 National Basketball Association games over fifteen years, paying special attention to the games' scores at halftime. It's not surprising that teams ahead at halftime won more games than teams that were behind. For example, a six-point halftime lead gives a team about an 80 percent probability of winning the game. However, Berger and Pope detected an exception to the rule: Teams that were behind *by just one point* were more likely to win. Indeed, being down by one at halftime was more advantageous than being up by one. Home teams with a one-point deficit at halftime won more than 58 percent of the time. Indeed, trailing by one point at halftime, weirdly, was equivalent to being *ahead* by two points.[20]

Berger and Pope then looked at ten years' worth of NCAA matchups, nearly 46,000 games in all, and found the same, though somewhat smaller, effect. "Being slightly behind [at halftime] significantly increases a team's chance of winning," they write. And when they examined the scoring patterns in greater detail, they found that the trailing teams scored a disproportionate number of their points immediately after the halftime break. They came out strong at the start of the second half.

Truckloads of sports data can reveal correlations, but they don't tell us anything definitive about causes. So Berger and Pope conducted some simple experiments to identify the mechanisms at work. They gathered participants and pitted each one against an opponent in another room in a contest to see who would bang out computer keystrokes more quickly. Those who scored higher than their opponents won a cash prize. The game had two short periods separated by a break. And it was during the break that experimenters treated their participants differently. They told some that they were far behind their opponent, some that they were a little behind, some that they were tied, and some that they were a little ahead.

The results? Three groups matched their first-half performance, but one did considerably better—the people who believed they were trailing by a little. "[M]erely telling people they were slightly behind an opponent led them to exert more effort," Berger and Pope write.[21]

In the second half of the 1982 finals, North Carolina came out blazing with an up-tempo offense and a swarming defense. Within four minutes, the Tar Heels had overcome their deficit and opened a three-point lead. But Georgetown and Ewing fought back, and the game seesawed its way into the final minutes. With thirty-two seconds left, Georgetown had moved to a 62–61 lead. Dean Smith called a time-out, his team down by one. North Carolina inbounded the ball, made seven passes near the top of the key, and then dished the ball to the weak side of the court, where a little-known freshman guard sunk a sixteen-foot jump shot to put the Tar Heels ahead. In the remaining seconds, the Hoyas floundered. And North Carolina's one-point halftime deficit became a one-point national championship victory.

The 1982 NCAA championship game became legendary in the annals of basketball. Dean Smith, John Thompson, and James Worthy would become three of only about 350 players, coaches, and other figures in the history of the game to earn plaques in the Naismith Memorial Basketball Hall of Fame. And that obscure freshman who hit the game winner was named Michael Jordan, whose basketball career worked out pretty well.

But for those of us interested in the psychology of midpoints, the most crucial moment came when Smith talked to his team when they were behind by one point. "We're in great shape," he told them. "I'd rather be in our shoes than theirs. We are exactly where we want to be."[22]

Midpoints are both a fact of life and a force of nature, but that doesn't make their effects inexorable. The best hope for turning a slump into a spark involves three steps.

First, be aware of midpoints. Don't let them remain invisible.

Second, use them to wake up rather than roll over—to utter an anxious "uh-oh" rather than a resigned "oh, no."

Third, at the midpoint, imagine that you're behind—but only by a little. That will spark your motivation and maybe help you win a national championship.

Time Hacker's Handbook

· CHAPTER 4 ·

FIVE WAYS TO REAWAKEN YOUR MOTIVATION DURING A MIDPOINT SLUMP

If you've reached the midpoint of a project or assignment, and the uh-oh effect hasn't kicked in, here are some straightforward, proven ways to dig yourself out of the slump:

1. Set interim goals.

To maintain motivation, and perhaps reignite it, break large projects into smaller steps. In one study that looked at losing weight, running a race, and accumulating enough frequent-flier miles for a free ticket, researchers found that people's motivation was strong at the beginning and end of the pursuit—but at the halfway mark became "stuck in the middle."[1] For instance, in the quest to amass 25,000 miles, people were more willing to work hard to accumulate miles when they had 4,000 or 21,000. When they had 12,000, though, diligence flagged. One solution is to get your mind to look at the middle in a different way. Instead of thinking about all 25,000 miles, set a subgoal at the 12,000-mile mark to accumulate 15,000 and make that your focus. In a race, whether literal or metaphorical, instead of imagining your distance from the finish line, concentrate on getting to the next mile marker.

2. Publicly commit to those interim goals.

Once you've set your subgoals, enlist the power of public commitment. We're far more likely to stick to a goal if we have someone holding us accountable. One way to surmount a slump is to tell someone else how and when you'll get something done. Suppose you're halfway through writing a thesis, or designing a curriculum, or crafting your organization's strategic plan. Send out a tweet or post to Facebook saying that you'll finish your current section by a certain date. Ask your followers to check in with you when that time comes. With so many people expecting you to deliver, you'll want to avoid public shame by reaching your subgoal.

3. Stop your sentence midway through.

Ernest Hemingway published fifteen books during his lifetime, and one of his favorite productivity techniques was one I've used myself (even to write this book). He often ended a writing session not at the end of a section or paragraph but smack in the middle of a sentence. That sense of incompletion lit a midpoint spark that helped him begin the following day with immediate momentum. One reason the Hemingway technique works is something called the Zeigarnik effect, our tendency to remember unfinished tasks better than finished ones.[2] When you're in the middle of a project, experiment by ending the day partway through a task with a clear next step. It might fuel your day-to-day motivation.

4. Don't break the chain (the Seinfeld technique).

Jerry Seinfeld makes a habit of writing every day. Not just the days when he feels inspired—every single damn day. To maintain focus, he prints a calendar with all 365 days of the year. He marks off each day he writes with a big red X. "After a few days, you'll have a chain," he told software developer Brad Isaac. "Just keep at it and the chain will grow longer every day. You'll like seeing that chain, especially when you get a few weeks under your belt. Your only job next is to not break the chain."[3] Imagine feeling the midpoint slump but then looking up at that string of thirty, fifty, or one hundred Xs. You, like Seinfeld, will rise to the occasion.

5. Picture one person your work will help.

To our midpoint-motivation murderer's row of Hemingway and Seinfeld, let's add Adam Grant, the Wharton professor and author of *Originals* and *Give and Take*. When he's confronted with tough tasks, he musters motivation by asking himself how what he's doing will benefit other people.[4] The slump of *How can I continue?* becomes the spark of *How can I help?* So if you're feeling stuck in the middle of a project, picture one person who'll benefit from your efforts. Dedicating your work to that person will deepen your dedication to your task.

ORGANIZE YOUR NEXT PROJECT WITH THE FORM-STORM-PERFORM METHOD

In the 1960s and 1970s, organizational psychologist Bruce Tuckman developed an influential theory of how groups move through time. Tuckman believed that all teams proceeded through four stages: forming, storming, norming, and performing. We can combine pieces of Tuckman's model with Gersick's research on team phases to create a three-phase structure for your next project.

Phase 1: Form and Storm.

When teams first come together, they often enjoy a period of maximal harmony and minimal conflict. Use those early moments to develop a shared vision, establish group values, and generate ideas. Eventually, though, conflict will break through the sunny skies. (That's Tuckman's "storm.") Some personalities may attempt to exert their influence and stifle quieter voices. Some people may contest their responsibilities and roles. As time passes, make sure all participants have a voice, that expectations are clear, and that all members can contribute.

Phase Two: The Midpoint.

For all the Sturm und Drang of phase one, your team probably hasn't accomplished much yet. That was Gersick's key insight. So use the midpoint—and the uh-oh effect it brings—to set direction and accelerate the pace. The University of Chicago's Ayelet Fishbach, whose work on Hanukkah candles I described earlier, has found that when team commitment to achieving a goal is high, it's best to emphasize the work that remains. But when team commitment is low, it's wiser to emphasize progress that has already been made even if it's not massive.[5] Figure out your own team's commitment and move accordingly. As you set the path, remember that teams generally become less open to new ideas and solutions after the midpoint.[6] However, they are also the most open to coaching.[7] So channel your inner Dean Smith, explain that you're a little behind, and galvanize action.

Phase Three: Perform.

At this point, team members are motivated, confident about achieving the goal, and generally able to work together with minimal friction. Keep the progress going but be wary of regressing back to the "storm" stage. Let's say you're part of a car-design team where different designers generally get along but are starting to become hostile to one another. To maintain optimal performance, ask your colleagues to step back, respect one another's roles, and reemphasize the shared vision they are moving toward. Be willing to shift tactics, but in this stage, direct your focus squarely on execution.

FIVE WAYS TO COMBAT A MIDLIFE SLUMP

Author and University of Houston professor Brené Brown offers a wonderful definition of "midlife." She says it's the period "when the Universe grabs your shoulders and tells you 'I'm not f—ing around, use the gifts you were given.'" Since most of us will someday contend with the U-curve of well-being, here are some ways to respond when the universe grabs your shoulders but you're not quite ready.

1. Prioritize your top goals (the Buffett technique).

As billionaires go, Warren Buffett seems like a pretty good guy. He's pledged his multibillion-dollar fortune to charity. He maintains a modest lifestyle. And he continues to work hard well into his eighties. But the Oracle of Omaha also turns out to be oracular in dealing with the midlife slump.

As legend has it, one day Buffett was talking with his private pilot, who was frustrated that he hadn't achieved all he'd hoped. Buffett prescribed a three-step remedy.

First, he said, write down your top twenty-five goals for the rest of your life.

Second, look at the list and circle your top five goals, those that are unquestionably your highest priority. That will give you two lists—one with your top five goals, the other with the next twenty.

Third, immediately start planning how to achieve those top five goals. And the other twenty? Get rid of them. Avoid them at all costs. Don't even look at them until you've achieved the top five, which might take a long time.

Doing a few important things well is far more likely to propel you out of the slump than a dozen half-assed and half-finished projects are.

2. Develop midcareer mentoring within your organization.

Most career mentorship happens when people are new to a field or business, and then disappears, fueled by the belief that we're fully established and no longer need guidance.

Hannes Schwandt of the University of Zurich says that's a mistake. He suggests providing formal, specific mentorship for employees *throughout* their career.[8] This has two benefits. First, it recognizes that the U-shaped curve of well-being is something most of us encounter. Talking openly about the slump can help us realize that it's fine to experience some midcareer ennui.

Second, more experienced employees can offer strategies for dealing with the slump. And peers can offer guidance to one another. What have people done to reinject purpose into their work? How have they built meaningful relationships in the office and beyond?

3. Mentally subtract positive events.

In the mathematics of midlife, sometimes subtraction is more powerful than addition. In 2008 four social psychologists borrowed from the movie *It's a Wonderful Life* to suggest a novel technique based on that idea.[9]

Begin by thinking about something positive in your life—the birth of a child, your marriage, a spectacular career achievement. Then list all the circumstances that made it possible—perhaps a seemingly insignificant decision of where to eat dinner one night or a class you decided to enroll in on a whim or the friend of a friend of a friend who happened to tell you about a job opening.

Next, write down all the events, circumstances, and decisions that might never have happened. What if you didn't go to that party or chose another class or skipped coffee with your cousin? Imagine your life without that chain of events and, more important, without that huge positive in your life.

Now return to the present and remind yourself that life did go your way. Consider the happy, beautiful randomness that brought that person or opportunity into your life. Breathe a sigh of relief. Shake your head at your good fortune. Be grateful. Your life may be more wonderful than you think.

4. Write yourself a few paragraphs of self-compassion.

We're often more compassionate toward others than we are toward ourselves. But the science of what's called "self-compassion" is showing that this bias can harm our well-being and undermine resilience.[10] That's why people who research this topic increasingly recommend practices like the following.

Start by identifying something about yourself that fills you with regret, shame, or disappointment. (Maybe you were fired from a job, flunked a class, undermined a relationship, ruined your finances.) Then write down some specifics about how it makes you feel.

Then, in two paragraphs, write yourself an e-mail expressing compassion or understanding for this element of your life. Imagine what someone who cares about you might say. He would likely be more forgiving than you. Indeed, University of Texas professor Kristin Neff suggests you write your letter "from the perspective of an unconditionally loving imaginary friend." But mix understanding with action. Add a few sentences on what changes you can make to your life and how you can improve in the future. A self-compassion letter operates like the converse corollary of the Golden Rule: It offers a way to treat yourself as you would others.

5. Wait.

Sometimes the best course of action is . . . inaction. Yes, that can feel agonizing, but no move can often be the right move. Slumps are normal, but they're also short-lived. Rising out of them is as natural as falling into them. Think of it as if it were a cold: It's a nuisance, but eventually it'll go away, and when it does, you'll barely remember it.

5.

ENDINGS

Marathons, Chocolates, and the Power of Poignancy

If you want a happy ending, that depends, of course, on where you stop your story.

—ORSON WELLES

Each year, more than half a million people in America run a marathon. After training for months, they rise early one weekend morning, lace up their shoes, and race 26.2 miles in one of the 1,100 marathons held annually in the United States. Elsewhere in the world, cities and regions host about 3,000 other marathons, which draw well over one million additional runners. Many of these participants, in the United States and across the globe, are running their very first marathon. By some estimates, about half the people in a typical marathon are first-timers.[1]

What compels these newbies to risk battered knees, twisted ankles, and the overconsumption of sports drinks? For Red Hong Yi, an artist in Australia, "a marathon was always one of those impossible things to do," she told me, so she decided to "give up my weekends and just go for it." She ran the 2015 Melbourne Marathon, her first, after training for six months. Jeremy Medding, who works in the diamond business in Tel Aviv and for whom the 2005 New York City Marathon was his first, told me that "there's always a goal we promise ourselves" and that a marathon was one box he hadn't ticked. Cindy Bishop, a lawyer in central Florida, said she ran her first marathon in 2009 "to change my life and reinvent myself." Andy Morozovsky, a zoologist turned biotech executive, ran the 2015 San Francisco Marathon even though he'd previously never run anywhere close to that distance. "I didn't plan to win it. I just planned to finish it," he told me. "I wanted to see what I could do."

Four people in four different professions living in four different parts of the world, all united by the common quest to run 26.2 miles. But something else links these runners and legions of other first-time marathoners.

Red Hong Yi ran her first marathon when she was twenty-nine years old. Jeremy Medding ran his when he was thirty-nine. Cindy Bishop ran her first marathon at age forty-nine, Andy Morozovsky at age fifty-nine.

All four of them were what social psychologists Adam Alter and Hal Hershfield call "9-enders," people in the last year of a life decade. They each pushed themselves to do something at ages twenty-nine, thirty-nine, forty-nine, and fifty-nine that they didn't do, didn't even consider, at ages twenty-eight, thirty-eight, forty-eight, and fifty-eight. Reaching the end of a decade somehow rattled their thinking and redirected their actions. Endings have that effect.

Like beginnings and midpoints, endings quietly steer what we do and how we do it. Indeed, endings of all kinds—of experiences, projects, semesters, negotiations, stages of life—shape our behavior in four predictable ways. They help us energize. They help us encode. They help us edit. And they help us elevate.

ENERGIZE: WHY WE KICK HARDER NEAR (SOME) FINISH LINES

Chronological decades have little material significance. To a biologist or physician, the physiological differences between, say, thirty-nine-year-old Fred and forty-year-old Fred aren't vast—probably not much different from those between Fred at thirty-eight and Fred at thirty-nine. Nor do our circumstances diverge wildly in years that end in nine compared with those that end in zero. Our life narratives often progress from segment to segment, akin to the chapters of a book. But the actual story doesn't abide by round numbers any more than novels do. After all, you wouldn't assess a book by its page numbers: "The one-hundred-sixties were super exciting, but the one-hundred-seventies were a little dull." Yet, when people near the end of the arbitrary marker of a decade, something awakens in their minds that alters their behavior.

For example, to run a marathon, participants must register with race organizers and include their age. Alter and Hershfield found that 9-enders are overrepresented among first-time marathoners by a whopping 48 percent. Across the entire life span, the age at which people were most likely to run their first marathon was twenty-nine. Twenty-nine-year-olds were about twice as likely to run a marathon as twenty-eight-year-olds or thirty-year-olds.

Meanwhile, first-time marathon participation declines in the early forties but spikes dramatically at age forty-nine. Someone who's forty-nine is about three times more likely to run a marathon than someone who's just a year older.

What's more, nearing the end of a decade seems to quicken a runner's pace. People who had run multiple marathons posted better times at age twenty-nine and thirty-nine than during the two years before or after those ages.[2]

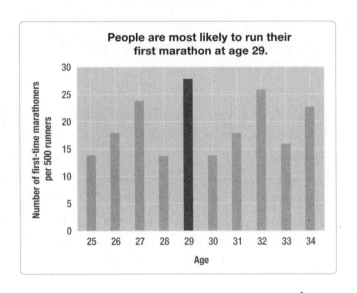

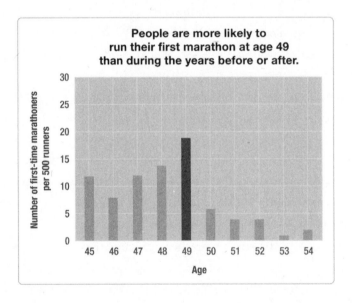

The energizing effect of the end of a decade doesn't make logical sense to marathon-running scientist Morozovsky. "Keeping track of our age? The Earth doesn't care. But people do, because we have short lives. We keep track to see how we're doing," he told me. "I wanted to accomplish this physical challenge before I hit sixty. I just did." For Yi, the Australian artist, the sight of that chronological mile marker roused her motivation. "As I was approaching the big three-oh, I had to really achieve something in my twenty-ninth year," she said. "I didn't want that last year just to slip by."

However, flipping life's odometer to a nine doesn't always trigger healthy behavior. Alter and Hershfield also discovered that "the suicide rate was higher among 9-enders than among people whose ages ended in any other digit." So, apparently, was the propensity of men to cheat on their wives. On the extramarital-affair website Ashley Madison, nearly one in eight men were twenty-nine, thirty-nine, forty-nine, or fifty-nine, about 18 percent higher than chance would predict.

What the end of the decade does seem to trigger, for good and for ill, is a reenergized pursuit of significance. As Alter and Hershfield explain:

> Because the approach of a new decade represents a salient boundary between life stages and functions as a marker of progress through the life span, and because life transitions tend to prompt changes in evaluations of the self, people are more apt to evaluate their lives as a chronological decade ends than they are at other times. 9-enders are particularly preoccupied with aging and meaningfulness, which is linked to a rise in behaviors that suggest a search for or crisis of meaning.[3]

Reaching the end also stirs us to act with greater urgency in other arenas. Take the National Football League. Each game lasts sixty minutes, two thirty-minute halves. In the ten years spanning the 2007–8 and 2016–17 seasons, according to STATS LLC, teams scored a total of 119,040 points. About 50.7 percent of those points came in the first half and about 49.3 percent in the second half—not much of a difference, especially considering that teams with leads late in the game often try not to score but run out the clock instead. But look a few statistical layers deeper, to the minute-by-minute scoring patterns, and the energizing effect of endings is apparent. During these seasons, teams scored a total of about 3,200 points in the final minute of the games, which was higher than almost all other one-minute game segments. But it was nothing compared to the nearly 7,900 points teams scored in the final minute of the first half. During the minute the half is ending, when the team that possesses the ball has every incentive to put points on the board, teams score well more than double what they score during any other minute of the game.[4]

Clark Hull, even though he was born forty years before the NFL's founding, would not have been surprised. Hull was a prominent American psychologist of the early twentieth century, one of the leading figures in behaviorism, which held that human beings behave not much differently from rats in a maze. In the early 1930s, Hull proposed what he called the "goal gradient hypothesis."[5] He built a long runway that he divided into equal sections. He placed food at every "finish line." Then he sent rats down the runway and timed how fast they ran in each section. He found that "animals in traversing a maze will move at a progressively more rapid pace as the goal is approached."[6] In other words, the closer the rats got to the vittles, the faster they ran. Hull's goal gradient hypothesis has held up far longer than most other behaviorist insights. At the beginning of a pursuit, we're generally more motivated by how far we've progressed; at the end, we're generally more energized by trying to close the small gap that remains.[7]

The motivating power of endings is one reason that deadlines are often, though not always, effective. For example, Kiva is a nonprofit organization that finances small low-interest or interest-free loans to micro-entrepreneurs. Prospective borrowers must complete a lengthy online application to be considered for a loan. Many of them begin the application but don't finish it. Kiva enlisted the Common Cents Lab, a behavioral research laboratory, to come up with a solution. Their suggestion: Impose an ending. Give people a specific deadline a few weeks away for completing the application. On one level, this idea seems idiotic. A deadline surely means that some people won't finish the application in time and therefore will be disqualified for the loan. But Kiva found that when it sent applicants a reminder message with a deadline, compared with a reminder message without a deadline, 24 percent more borrowers completed the application.[8] Likewise, in other studies, people given a hard deadline—a date and time—are more likely to sign up to be organ donors than those for whom the choice is open-ended.[9] People with a gift certificate valid for two weeks are three times more likely to redeem it than people with the same gift certificate valid for two months.[10] Negotiators with a deadline are far more likely to reach an agreement than those without a deadline—and that agreement comes disproportionately at the very end of the allotted time.[11]

Think of this phenomenon as a first cousin of the fresh start effect—the fast finish effect. When we near the end, we kick a little harder.

To be sure, the effect is not uniform or entirely positive. For instance, as we close in on a finish line, having multiple ways to cross it can slow our progress.[12] Deadlines, especially for creative tasks, can sometimes reduce intrinsic motivation and flatten creativity.[13] And while imposing a finite end to negotiations—for labor-management contracts or even peace agreements—can often speed a resolution, that doesn't always lead to the best or most enduring results.[14]

However, as with Clark Hull's rats, being able to sniff the finish line—whether it offers a hunk of cheese or a slice of meaning—can invigorate us to move faster.

Red Hong Yi, now thirty-one, continues to run for exercise, although she hasn't attempted a second marathon or even contemplated running one in the next few years. "Maybe I can do it on my thirty-ninth birthday," she says.

ENCODE: JIMMY, JIM, AND THE GOOD LIFE

On February 8, 1931, Mildred Marie Wilson of Marion, Indiana, gave birth to what would be her only child, a baby boy that she and her husband named James and called Jimmy. Jimmy enjoyed a happy, if tumultuous, childhood. His family moved from northern Indiana to Southern California when he began elementary school. But a few years later, his mother died suddenly of cancer—and Jimmy's bereft father sent him back to Indiana to live with relatives. The rest of his young life was pleasant and steady in a midwestern way—church, sports teams, debate club. When he graduated high school, he moved back to Southern California for college, where he caught the movie bug, and in 1951, just shy of turning twenty, he dropped out of UCLA to pursue an acting career.

Then this ordinary story took an extraordinary turn.

Jimmy quickly landed a few commercials and minor television roles. And the year he turned twenty-three, one of the era's most famous directors cast him in the film adaptation of a John Steinbeck novel. The movie became a hit; Jimmy was nominated for an Oscar. That same year, he landed the lead role in an even more prominent movie; it earned him another Oscar nomination. In a blink, at an impossibly young age, he became an impossibly huge Hollywood star. Then, about four months shy of his twenty-fifth birthday, Jimmy, whose full name was James Byron Dean, died in an auto accident.

Stop for a moment and ponder this question: Taking Jimmy's life as a whole, how desirable do you think it was? On a 1-to-9 scale, with 1 being the most undesirable life and 9 being the most desirable life, what number would you assign?

Now consider a hypothetical. Imagine that Jimmy had lived a few more decades but that he never achieved the professional success of his early twenties. He didn't spiral into homelessness or drug addiction. His career didn't implode. His star just fell from its empyrean heights. Maybe he did a TV sitcom or two and won a few smaller parts in less successful films before he died, say, in his midfifties. How would you rate his life now?

When researchers have studied scenarios like these, they've uncovered something strange. People tend to rate lives like the first scenario (a short life that ends on an upswing) more highly than those like the second (a longer life that ends on a downswing). Considered in purely utilitarian terms, this conclusion is bizarre. After all, in the hypothetical, Jimmy lives thirty years longer! And those extra years aren't choked with misery; they're simply less spectacular than the early ones. The *cumulative* amount of positivity of that longer life (which still includes those early years as a star) is indisputably higher.

"The suggestion that adding mildly pleasant years to a very positive life does not enhance, but decreases, perceptions of the quality of life is counterintuitive," write social scientists Ed Diener, Derrick Wirtz, and Shigehiro Oishi. "We label this the James Dean Effect because a life that is short but intensely exciting, such as the storied life led by the actor James Dean, is seen as most positive."[15]

The James Dean effect is another example of how endings alter our perception. They help us encode—that is, to evaluate and record—entire experiences. You might have heard of the "peak-end rule." Formulated in the early 1990s by Daniel Kahneman and colleagues including Don Redelmeier and Barbara Fredrickson, who studied patient experiences during colonoscopies and other unpleasant experiences, the rule says that when we remember an event we assign the greatest weight to its most intense moment (the peak) and how it culminates (the end).[16] So a shorter colonoscopy in which the final moments are painful is remembered as being worse than a longer colonoscopy that happens to end less unpleasantly even if the latter procedure delivers substantially more total pain.[17] We downplay how long an episode lasts—Kahneman calls it "duration neglect"—and magnify what happens at the end.[18]

The encoding power of endings shapes many of our opinions and subsequent decisions. For instance, several studies show that we often evaluate the quality of meals, movies, and vacations not by the full experience but by certain moments, especially the end.[19] So when we share our evaluations with others—in conversations or in a TripAdvisor review—much of what we're conveying is our reaction to the conclusion. (Look at Yelp reviews of restaurants, for example, and notice how many of the reviews describe how the meal ended—an unexpected farewell treat, a check with an error, a server chasing after diners to return an item left behind.) Endings also affect more consequential choices. For example, when Americans vote for president, they tell pollsters they intend to decide based on the full four years of an expiring presidential term. But research shows voters decide based on the *election year* economy—the culmination of a four-year sequence, not its totality. This "end heuristic," political scientists argue, leads to "myopic voting" and, perhaps as a result, myopic policies.[20]

The encoding effects of endings are especially strong when it comes to our idea of what constitutes a moral life. Three Yale researchers set up an experiment using different versions of a short biography of a fictitious character they called Jim. In all the versions, Jim was the CEO of a company, but the researchers varied the trajectories of Jim's life. In some cases, he was a nasty guy who underpaid his employees, denied them health care benefits, and never gave to charities—behavior that lasted for three decades. But late in his career, close to retirement, he turned generous. He raised pay, shared profits, and "started donating large amounts of money to various charities around the community"—only to die suddenly of a surprise heart attack six months after he turned benevolent. In other scenarios, Jim moved in the opposite direction. For several decades, he was a kind and generous CEO—"putting the wellbeing of his employees ahead of his own financial interests" and donating large sums to local charities. But as he neared retirement, he "dramatically altered his behavior." He cut salaries, began taking most of the profits for himself, and ceased his charitable giving—only to die suddenly of a surprise heart attack six months later.[21]

The researchers gave half their participants the bad-guy-to-good-guy bio and half the good-guy-to-bad-guy bio, and asked both groups to evaluate Jim's overall moral character. Across multiple versions of the study, people assessed Jim's morality based largely on how he behaved at the *end* of his life. Indeed, they evaluated a life with twenty-nine years of treachery and six months of goodness the same as a life with twenty-nine years of goodness and six months of treachery. "[P]eople are willing to override a relatively long period of one kind of behavior with a relatively short period of another kind just because it occurred at the end of one's life."[22] This "end of life bias," as the researchers call it, suggests that we believe people's true selves are revealed at the end—even if their death is unexpected and the bulk of their lives evinced a far different self.

Endings help us encode—to register, rate, and recall experiences. But in so doing, they can distort our perceptions and obscure the bigger picture. Of the four ways that endings influence our behavior, encoding is the one that should make us most wary.

EDIT: WHY LESS IS MORE—
ESPECIALLY NEAR THE END

Our lives are not always dramatic, but they can unfold like a three-act drama. Act one: the launch. We move from childhood to young adulthood, then eagerly set out to establish a foothold in the world. Act two: Harsh realities descend. We scramble to earn a living, maybe find a mate and start a family. We advance, suffer setbacks, mix triumph with disappointment. Act three: the bittersweet culmination. Maybe we've achieved something. Maybe we have people who love us. Yet the denouement is near, the curtain about to fall.

The other characters—our collection of friends and family—appear throughout the drama. But as Tammy English of Washington University in St. Louis and Laura Carstensen of Stanford University discovered, their time onstage varies from act to act. English and Carstensen looked at ten years of data on people aged eighteen to ninety-three to determine how their social networks and friendships shifted over the three acts of life. (The researchers themselves didn't divide the ages by acts. I'm layering that notion on top of their data to

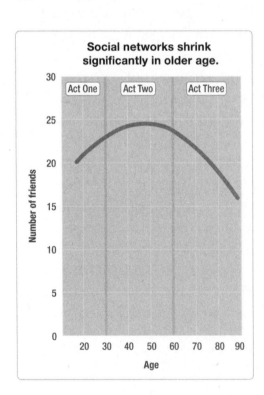

Social networks shrink significantly in older age.

Number of friends

Age

illuminate a point.) As you can see in the chart, when people reached about the age of sixty, their number of friendships plunged and the size of their social network shrank.

This makes intuitive sense. When we leave the workforce, we can lose connections and friends that once enriched our daily lives. When our kids depart home and enter their own act twos, we often see them less and miss them more. When we reach our sixties and seventies, our contemporaries begin dying, extinguishing lifelong relationships and leaving us with fewer peers. The data confirm what we've long suspected: Act three is full of pathos. Old age can be lonely and isolating. It's a sad story.

But it's not a true story.

Yes, older people have much smaller social networks than when they were younger. But the reason isn't loneliness or isolation. The reason is both more surprising and more affirming. It's what we *choose*. As we get older, when we become conscious of the ultimate ending, we *edit* our friends.

English and Carstensen asked people to draw their social networks and place themselves in the center surrounded by three concentric circles. The inner circle was for "people you feel very close to, so close that it

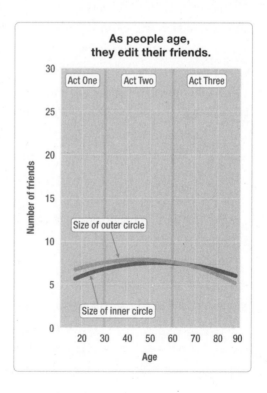

would be hard to imagine life without them." The middle circle was for people who were still important but less close than the inner circle. In the outer circle were people the respondents felt a little less close to than the middle circle. Look at the chart that shows the size of the inner and outer circles over time.

A bit after age sixty, the outer circle begins to decline, but the inner circle remains about the same size. Then in the mid to late sixties, the number of people in the inner circle edges ahead of those in the outer circle.

"As participants aged, there was a decline in the number of peripheral partners . . . but great stability in the number of close social partners into late life," English and Carstensen found. However, the outer and middle circle friends didn't quietly creep offstage in act three. "They were actively eliminated," the researchers say. Older people have fewer total friends not because of circumstance but because they've begun a process of "active pruning, that is, removing peripheral partners with whom interactions are less emotionally meaningful."[23]

Carstensen began developing this idea in 1999 when she (and two of her former students) published a paper titled "Taking Time Seriously." "As people move through life," she wrote, "they become increasingly aware that time is in some sense 'running out.' More social contacts feel superficial—trivial—in contrast to the ever-deepening ties of existing close relationships. It becomes increasingly important to make the 'right' choice, not to waste time on gradually diminishing future payoffs."[24]

Carstensen called her theory "socioemotional selectivity." She argued that our perspective on time shapes the orientation of our lives and therefore the goals we pursue. When time is expansive and open-ended, as it is in acts one and two of our lives, we orient to the future and pursue "knowledge-related goals." We form social networks that are wide and loose, hoping to gather information and forge relationships that can help us in the future. But as the horizon nears, when the future is shorter than the past, our perspective changes. While many believe that older people pine for yesteryear, Carstensen's body of work shows something else. "The primary age difference in time orientation concerns not the past but the present," she wrote.[25]

When time is constrained and limited, as it is in act three, we attune to the now. We pursue different goals—emotional satisfaction, an appreciation for life, a sense of meaning. And these updated goals make people "highly selective in their choice of social partners" and prompt them to "systematically hone their social networks." We edit our relationships. We omit needless people. We choose to spend our remaining years with networks that are small, tight, and populated with those who satisfy higher needs.[26]

Moreover, what spurs editing isn't aging per se, Carstensen found, but endings of any sort. For example, when she compared college seniors with new college students, students in their final year displayed the same kind of social-network pruning as their seventy-something grandparents. When people are about to switch jobs or move to a new city, they edit their immediate social networks because their time in that setting is ending. Even political transitions have this effect. In a study of people in Hong Kong four months before the territory's handover from Great Britain to the People's Republic of China in 1997, both young people and older folks narrowed their circles of friends.

Just as intriguing, the converse is also true: *Expanding* people's time horizons *arrests* their editing behavior. Carstensen conducted an experiment in which she asked people to "imagine that they had just received a telephone call from their physician, who had informed them of a new medical breakthrough that would likely add 20 years to their life." Under these conditions, older people were no more likely than younger ones to prune their social networks.[27]

Yet, when endings become salient—whenever we enter an act three of any kind—we sharpen our existential red pencils and scratch out anyone or anything nonessential. Well before the curtain falls, we edit.

ELEVATE: GOOD NEWS, BAD NEWS, AND HAPPY ENDINGS

"I've got some good news and some bad news."

You've undoubtedly said that before. Whether you're a parent, a teacher, a doctor, or a writer trying to explain a missed deadline, you had to deliver information—some of it positive, some of it not—and opened with this two-headed approach.

But which piece of information should you introduce first? Should the good news precede the bad? Or should the happy follow the sad?

As someone who finds himself delivering mixed news more often than he should or wants to, I've always led with the positive. My instinct has been to spread a downy duvet of good feeling to cushion the coming hammerblow.

My instinct, alas, has been dead wrong.

To understand why, let's switch perspectives—from me to you. Suppose you're on the *receiving* end of my mixed news, and after my "I've got some good news and some bad news" windup, I append a question: "Which would you like to hear first?"

Think about that for a moment.

Chances are, you opted to hear the bad news first. Several studies over several decades have found that roughly four out of five people "prefer to begin with a loss or negative outcome and ultimately end with a gain or positive outcome, rather than the reverse."[28] Our preference, whether we're a patient getting test results or a student awaiting a midsemester evaluation, is clear: bad news first, good news last.

But as news givers, we often do the reverse. Delivering that harsh performance review feels unsettling, so we prefer to ease into it, to demonstrate our kind intentions and caring nature by offering a few spoonfuls of sugar before administering the bitter medicine. Sure, we know that *we* like to hear the bad news first. But somehow we don't understand that the person sitting across the desk, wincing at our two-headed intro, feels the same. She'd rather get the grimness out of the way and end the encounter on a more redeeming note. As two of the researchers who've studied this issue say, "Our findings suggest that the doctors, teachers, and partners . . . might do a poor job of giving good and bad news because they forget for a moment how they want to hear news when they are patients, students, and spouses."[29]

We blunder—I blunder—because we fail to understand the final principle of endings: Given a choice, human beings prefer endings that elevate. The science of timing has found—repeatedly—what seems to be an innate preference for happy endings.[30] We favor sequences of events that rise rather than fall, that improve rather than deteriorate, that lift us up rather than bring us down. And simply knowing this inclination can help us understand our own behavior and improve our interactions with others.

For example, social psychologists Ed O'Brien and Phoebe Ellsworth of the University of Michigan wanted to see how endings shaped people's judgment. So they packed a bag full of Hershey's Kisses and headed to a busy area of the Ann Arbor campus. They set up a table and told students they were conducting a taste test of some new varieties of Kisses that contained local ingredients.

People sidled up to the table, and a research assistant, who didn't know what O'Brien and Ellsworth were measuring, pulled a chocolate out of the bag and asked a participant to taste it and rate it on a 0-to-10 scale.

Then the research assistant said, "Here is your next chocolate," gave the participant another candy, and asked her to rate that one. Then the experimenter and her participant did the same thing again for three more chocolates, bringing the total number of candies to five. (The tasters never knew how many total chocolates they would be sampling.)

The crux of the experiment came just before people tasted the fifth chocolate. To half the participants, the research assistant said, "Here is your next chocolate." But to the other half of the group, she said, "Here is your last chocolate."

The people informed that the fifth chocolate was the last—that the supposed taste test was now ending—reported liking that chocolate much more than the people who knew it was simply next. In fact, people informed that a chocolate was last liked it significantly more than any other chocolate they'd sampled. They chose chocolate number five as their favorite chocolate 64 percent of the time (compared with the "next" group, which chose that chocolate as their favorite 22 percent of the time). "Participants who knew they were eating the final chocolate of a taste test enjoyed it more, preferred it to other chocolates, and rated the overall experience as more enjoyable than other participants who thought they were just eating one more chocolate in a series."[31]

Screenwriters understand the importance of endings that elevate, but they also know that the very best endings are not always happy in the traditional sense. Often, like a final chocolate, they're bittersweet. "Anyone can deliver a happy ending—just give the characters everything they want," says screenplay guru Robert McKee. "An artist gives us the emotion he's promised . . . but with a rush of unexpected insight."[32] That often comes when the main character finally understands an emotionally complex truth. John August, who wrote the screenplay for *Charlie and the Chocolate Factory* and other films, argues that this more sophisticated form of elevation is the secret to the success of Pixar films such as *Up*, *Cars*, and the *Toy Story* trilogy.

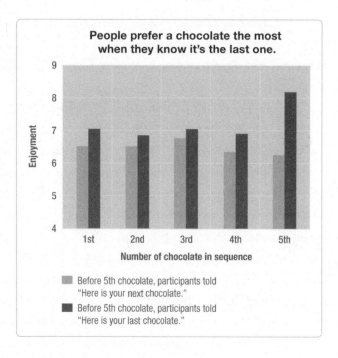

People prefer a chocolate the most
when they know it's the last one.

Enjoyment

Number of chocolate in sequence

■ Before 5th chocolate, participants told
"Here is your next chocolate."

■ Before 5th chocolate, participants told
"Here is your last chocolate."

"Every Pixar movie has its protagonist achieving the goal he wants only to realize it is not what the protagonist needs. Typically, this leads the protagonist to let go of what he wants (a house, the Piston Cup, Andy) to get what he needs (a true yet unlikely companion; real friends; a lifetime together with friends)."[33] Such emotional complexity turns out to be central to the most elevated endings.

Hal Hershfield, one of the 9-ender researchers I mentioned earlier in the chapter, and Laura Carstensen teamed up with two other scholars to explore what makes endings meaningful. In one of their studies, the researchers approached Stanford seniors on graduation day to survey them about how they felt. To one group, they gave the following instructions: "Keeping in mind your current experiences, please rate the degree to which you feel each of the following emotions," and then gave them a list of nineteen emotions. To the other group, they added one sentence to the instructions to raise the significance that something was ending: "As a graduating senior, today is the last day that you will be a student at Stanford. Keeping that in mind, please rate the degree to which you feel each of the following emotions."[34]

The researchers found that at the core of meaningful endings is one of the most complex emotions humans experience: poignancy, a mix of happiness and sadness. For graduates and everyone else, the most powerful endings deliver poignancy because poignancy delivers significance. One reason we overlook poignancy is that it operates by an upside-down form of emotional physics. Adding a small component of sadness to an otherwise happy moment *elevates* that moment rather than diminishes it. "Poignancy," the researchers write, "seems to be particular to the experience of endings." The best endings don't leave us happy. Instead, they produce something richer—a rush of unexpected insight, a fleeting moment of transcendence, the possibility that by discarding what we wanted we've gotten what we need.

Endings offer good news and bad news about our behavior and judgment. I'll give you the bad news first, of course. Endings help us encode, but they can sometimes twist our memory and cloud our perception by overweighting final moments and neglecting the totality.

But endings can also be a positive force. They can help energize us to reach a goal. They can help us edit the nonessential from our lives. And they can help us elevate—not through the simple pursuit of happiness but through the more complex power of poignancy. Closings, conclusions, and culminations reveal something essential about the human condition: In the end, we seek meaning.

Time Hacker's Handbook

• CHAPTER 5 •

READ LAST LINES

"In the late summer of that year we lived in a house in a village that looked across the river and the plain to the mountains."

The literary among you might recognize these words as the first sentence of Ernest Hemingway's *A Farewell to Arms*. In literature, opening lines bear a mighty burden. They must hook the reader and lure her into the book. That's why opening lines are heavily scrutinized and long remembered.

(Don't believe me? Then call me Ishmael.)

But what about last lines? The final words of a work are just as important and deserve comparable reverence. Last lines can elevate and encode—by encapsulating a theme, resolving a question, leaving the story lingering in the reader's head. Hemingway said that he rewrote the ending to *A Farewell to Arms* no fewer than thirty-nine times.

One easy way to appreciate the power of endings and improve your own ability to create them: Take some of your favorite books off the shelf and flip to the end. Read the last line. Read it again. Ponder it for a moment. Maybe even memorize it.

Here are some of my favorites to get you started:

"The creatures outside looked from pig to man, and from man to pig, and from pig to man again; but already it was impossible to say which was which."

—*Animal Farm*, George Orwell

"'It isn't fair, it isn't right,' Mrs. Hutchinson screamed, and then they were upon her."

—"The Lottery," Shirley Jackson

"For now he knew what Shalimar knew: If you surrendered to the air, you could ride it."

—*Song of Solomon*, Toni Morrison

"In a place far away from anyone or anywhere, I drifted off for a moment."

—*The Wind-Up Bird Chronicle*, Haruki Murakami

"So we beat on, boats against the current, borne back ceaselessly into the past."

—*The Great Gatsby*, F. Scott Fitzgerald

And that last sentence of *A Farewell to Arms*—the one Hemingway finally settled on? "After a while I went out and left the hospital and walked back to the hotel in the rain."

WHEN TO QUIT A JOB: A GUIDE

Many "when" decisions involve endings. And one of the biggest is when to leave a job that just isn't working out. That's a big step, a risky move, and not always a choice for some people. But if you're contemplating this option, here are five questions to help you decide.

If your answer to two or more of these is no, it might be time to craft an end.

1. Do you want to be in this job on your next work anniversary?

People are most likely to leave a job on their one-year work anniversary. The second most likely time? Their two-year anniversary. The third? Their three-year anniversary.[1] You get the idea. If you dread the thought of being at your job on your next work anniversary, start looking now. You'll be better prepared when the time comes.

2. Is your current job both demanding and in your control?

The most fulfilling jobs share a common trait: They prod us to work at our highest level but in a way that we, not someone else, control. Jobs that are demanding but don't offer autonomy burn us out. Jobs that offer autonomy but little challenge bore us. (And jobs that are neither demanding nor in our control are the worst of all.) If your job doesn't provide both challenge and autonomy, and there's nothing you can do to make things better, consider a move.

3. Does your boss allow you to do your best work?

In his excellent book *Good Boss, Bad Boss: How to Be the Best . . . and Learn from the Worst*, Stanford Graduate School of Business professor Robert Sutton explains the qualities that make someone worth working for. If your boss has your back, takes responsibility instead of blaming others, encourages your efforts but also gets out of your way, and displays a sense of humor rather than a raging temper, you're probably in a good place.[2] If your boss is the opposite, watch out—and maybe get out.

4. Are you outside the three- to five-year salary bump window?

One of the best ways to boost your pay is to switch organizations. And the best time to do that is often three to five years after you've started. ADP, the massive human resources management company, found that this period represents the sweet spot for pay increases.[3] Fewer than three years might be too little time to develop the most marketable skills. More than five years is when employees start becoming tied to their company and moving up its leadership ranks, which makes it more difficult to start somewhere else.

5. Does your daily work align with your long-term goals?

Ample research from many countries shows that when your individual goals align with those of your organization, you're happier and more productive.[4] So take a moment and list your top two or three goals for the next five years and ten years. If your current employer can help you reach them, great. If not, think about an ending.

WHEN TO QUIT A MARRIAGE: A HEDGE

When should you get divorced? This kind of ending is too fraught, the research too sprawling, the circumstances of people's lives too varied to offer a definitive answer. But some research indicates when your *spouse* might make the move.

Julie Brines and Brian Serafini analyzed fourteen years of divorce filings in the state of Washington and detected a distinct seasonal rhythm. Divorce filings spiked in the months of March and August, a pattern that they later found in four other states and that gave rise to a chart, shown on the next page, that resembles the Bat-Signal.[5]

The reasons for the two monthly peaks aren't clear. But Brines and others speculate that the twin peaks may be forged by domestic rituals and family calendars. "The high season for divorce attorneys

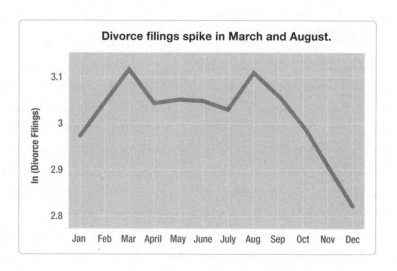

Divorce filings spike in March and August.

is January and February, when the holidays are over and people can finally stop pretending to be happy," says *Bloomberg Businessweek*.[6] Over the winter holidays, spouses often give a marriage one last try. But when the festivities end and disillusionment descends, they visit a divorce lawyer. Since contested divorces require some work, the papers aren't filed until four to six weeks later, which explains the March burst. The same thing might happen at the end of the school year. Parents keep it together for the kids. But once school is out, they head to the lawyer's office in June and July, resulting in another filing spike in August. Consider yourself warned.

FOUR AREAS WHERE YOU CAN CREATE BETTER ENDINGS

If we're conscious of the power of closing moments and our ability to shape them, we can craft more memorable and meaningful endings in many realms of life. Here are four ideas:

The workday

When the workday ends, many of us want to tear away—to pick up children, race home to prepare dinner, or just beeline to the nearest bar. But the science of endings suggests that instead of fleeing we're better off reserving the final five minutes of work for a few small deliberate actions that bring the day to a fulfilling close. Begin by taking two or three minutes to write down what you accomplished since the morning. Making progress is the single largest day-to-day motivator on the job.[7] But without tracking our "dones," we often don't know whether we're progressing. Ending the day by recording what you've achieved can encode the entire day more positively. (Testimonial: I've been doing this for four years and I swear by the practice. On good days, the exercise delivers feelings of completion; on bad days, it often shows me I got more done than I suspected.)

Now use the other two or three minutes to lay out your plan for the following day. This will help close the door on today and energize you for tomorrow.

Bonus: If you've got an extra minute left, send someone—anyone—a thank-you e-mail. I mentioned in chapter 2 that gratitude is a powerful restorative. It's an equally powerful form of elevation.

The semester or school year

At the end of a school term, many students feel a sense of relief. But with a little thought and planning, they can also experience a sense of elevation. That's why some inspired teachers are using endings as meaning makers. For example, Anthony Gonzalez, an economics teacher at Nazareth Academy outside of Chicago, has his seniors write a letter to themselves—which he mails to them five years later. "In it, they include wisdom from high school, guesses on careers, pay, what adventures they hope to go on, stock prices, and so on. It's a very cool opportunity for them to reflect." And it's a good way for Gonzalez to reconnect with them when they're twenty-three and high school is a distant memory.

At North High School in Des Moines, Iowa, choir teacher Vanessa Brady enlists her husband, Justin, on the last day of school to bring in griddles, butter, syrup, and his homemade pancake batter for an end-of-the-year Pancake Day.

For the last class of a term, Alecia Jioeva, who teaches at Lomonosov Moscow State University in Russia, takes her students to a small restaurant where they offer toasts to one another.

At the beginning of the school year, Beth Pandolpho, a language arts teacher at West Windsor–Plainsboro High School North in New Jersey, asks her students to write six-word memoirs that she hangs on a clothesline stretched around the perimeter of the classroom. At the end of the year, students write another six-word memoir. They read the earlier memoir aloud, remove it from the clothesline, and then read the new one. "To me," Pandolpho says, "it feels a little bit like bringing our time together full circle."

A vacation

How a vacation ends shapes the stories we later tell about the experience. As University of British Columbia psychologist Elizabeth Dunn explained to *New York* magazine, "[T]he very end of an experience seems to disproportionately affect our memory of it," which means that "going out with a bang, going on the hot air balloon or whatever on the last day of the trip, could . . . be a good strategy for maximizing reminiscence."[8] As you plan your next vacation, you needn't save all the best for last. But you'll enjoy the vacation more, both in the moment and in retrospect, if you consciously create an elevating final experience.

A purchase

For all the words scratched and bellowed about the importance of customer service, we've generally given short shrift to the end of encounters with customers and clients. Yes, some restaurants present guests with free chocolates when servers bring the check. And, yes, at Nordstrom stores, sales associates famously walk out from behind the counter to personally hand customers the purchase they've just made. But imagine if more organizations treated endings with greater respect and creativity. For example, what if at the end of the meal in which the guests have spent above a certain amount, restaurants handed the table a card asking the group to select one of three charities that the restaurant will make a small donation to in their name? Or what if someone at a store who's made a major purchase—a computer, an appliance, an expensive item of clothing—departs the establishment past a line of employees saying, "Thank you," and giving that customer a round of applause?

PART THREE.

SYNCHING AND THINKING

6.

SYNCHING FAST AND SLOW

The Secrets of Group Timing

That is happiness; to be dissolved into
something complete and great.

—WILLA CATHER, *My Ántonia*

On a muggy February morning, as what passes for sunshine glints off giant billboards advertising 50 percent discounts on wedding clothes, India's largest city is coming to life. Here in Mumbai, the tang of smoke hangs in the air. Cars, trucks, and auto-rickshaws clog the roads, honking like embittered geese. Office workers in slacks and saris stream through alleys and wash onto commuter trains. And Ahilu Adhav, age forty, adjusts his white cap and jumps on his bicycle to begin his rounds.

Adhav pedals through Mumbai's Vile Parle (pronounced *VEE-luh PAR-lay*) neighborhood, past street vendors selling everything from fresh cabbage to packaged socks, and steers toward the front of a small apartment building. He hops off the bike—the ability to quickly dismount moving vehicles is one of Adhav's many skills—strides into the building, and rides the elevator to the third-floor apartment of the Turakhia family.

It's 9:15 a.m. He presses the buzzer once, then twice. The door opens. After a quick apology for making him wait, Riyankaa Turakhia hands Adhav a maroon canvas bag about the size of a gallon of milk. Inside the bag is a cylindrical stack of four metal containers. Inside those containers, called tiffins, is her husband's lunch—cauliflower, yellow dahl, rice, and roti. In three and a half hours, this home-cooked lunch will appear on her husband's desk in downtown Mumbai, about thirty kilometers (nineteen miles) away. And in about seven hours, the canvas bag and its empty tiffins will reappear at this same door.

Adhav is a *dabbawala*. (*Dabba* is the Hindi word for those metal tiffin boxes, *wala* is an amalgam of "doer" and "merchant.") During the first sixty-eight minutes of his Monday, he will collect fifteen such lunches, tying each bag to the handlebars or the rear of his bike. Then, coordinating with a team of a dozen other dabbawalas who've collected their own bags elsewhere in this sprawling neighborhood of about half a million people, he will sort the lunches, hoist twenty of them on his back, board the luggage compartment of a commuter train, and deliver the lunches to shops and offices in the business districts of the city.

He's not alone: About 5,000 dabbawalas work in Mumbai. Each day they deliver more than 200,000 lunches. They do this six times a week nearly every week of the year—with an accuracy that rivals FedEx and UPS.

"In today's world, we're very health conscious," Turakhia tells me at Adhav's first stop. "We crave homemade food. And these guys do an excellent job of delivering the dabba to the right place at exactly the right time." Her husband, who works for a brokerage firm, leaves for the office at 7 a.m., too early for anyone to prepare a proper lunch. But the dabbawalas offer the family time and peace of mind. "They're very, very coordinated and synchronized," Turakhia says. In the five years she's enlisted Adhav and his crew, for a fee affordable to most

Dabbawala Ahilu Adhav fastens a lunch to the back of his bicycle.

middle-class urban families (about $12 per month), they've misdelivered the lunch or delivered it late exactly zero times.

What the dabbawalas manage to do every day verges on preposterous. Mumbai operates with a twenty-four-hour full-tilt intensity, a move-or-be-mowed-down ethos that makes Manhattan seem like a fishing village. Mumbai is not just one of the largest cities in the world; it is also one of the most densely populated. The sheer shoulder-to-shoulder humanity of the city itself—12 million citizens crammed into an area one-fifth the size of Rhode Island—gives it a throbbing, anarchic intensity. "A city in heat," journalist Suketu Mehta calls it.[1] Yet the dabbawalas somehow haul home-cooked meals in canvas bags through the chaos of Mumbai with military precision and punctuality.

More impressive, the dabbawalas are so deeply in synch with one another, so finely attuned to the tempo of their task, that they pull off the feat—200,000 lunch deliveries every day—without any technology beyond bicycles and trains.

No smartphones. No scanners. No bar codes. No GPS.

And no mistakes.

Human beings rarely go it alone. Much of what we do—at work, at school, and at home—we do in concert with other people. Our ability to survive, even to *live*, depends on our capacity to coordinate with others in and across time. Yes, individual timing—managing our beginnings, midpoints, and endings—is crucial. But group timing is just as important, and what lies at its heart is crucial for us to know.

Consider a patient wheeled into an emergency room with a serious heart attack. Whether that patient lives or dies depends on how well coordinated the medical professionals are—whether they can deftly synchronize their activities while the clock, and perhaps the patient's life, ticks away.

Or take less dire circumstances that require group timing. Software engineers who work on different continents in different time zones to ship a product by a certain date. Event planners who coordinate multiple crews of technicians, hospitality workers, and presenters so that a three-day conference can unfold on time and free of calamities. Political candidates who organize campaign volunteers to canvass neighborhoods, register voters, and distribute yard signs before Election Day. Schoolteachers who marshal sixty students on and off a bus and through a museum during a field trip. Sports teams. Marching bands. Shipping companies. Factories. Restaurants. All require individuals to work in tempo, to synchronize their actions with others, to move to a common beat and toward a common goal.

The breakthrough that most enabled us to do these things came in the late 1500s, when Galileo Galilei was a nineteen-year-old medical student at the University of Pisa. Inspired by a swinging chandelier, Galileo conducted a few makeshift experiments on pendulums. He discovered that what most affected a pendulum's motion was the length of its string—and that for any given length of string a pendulum always took the same amount of time to make one full swing. That periodicity, he concluded, made pendulums ideal timekeepers. Galileo's insight led to the invention of pendulum clocks a few decades later. And pendulum clocks, in turn, produced something that we don't realize is a relatively new concept: "the time."

Imagine life without even a rough consensus on what time it is. You'd find a way to manage. But it would be cumbersome and inefficient in ways we can scarcely fathom today. How would you know when to make a delivery, expect a bus, or take your kid to the dentist? Pendulum clocks, which were far more accurate than their predecessors, remade civilization by allowing people to synchronize their actions. Public clocks appeared in town squares and began establishing a single standard of time. Two o'clock for me became two o'clock for you. And this notion of public time—"the time"—greased the wheels of commerce and lubricated social interaction. Before long, local time standardization became regional, and regional standardization became national, giving rise to predictable schedules and the 5:16 p.m. train to Poughkeepsie.[2]

This ability to synchronize our actions with others, liberated by the cascade Galileo set off a few centuries ago, has been critical to human progress. Yet a consensus about what the clock says is only the first ingredient. Groups that depend on synchronization for success—choirs, rowing teams, and those Mumbai dabbawalas—abide by three principles of group timing. An external standard sets the pace. A sense of belonging helps individuals cohere. And synchronization both requires and heightens well-being.

Put another way, groups must synchronize on three levels—to the boss, to the tribe, and to the heart.

THE CHOIRMASTER, THE COXSWAIN, AND THE CLOCK: SYNCHING TO THE BOSS

David Simmons is the same height as Ahilu Adhav, but the resemblance dissolves where the tape measure ends. Simmons is white, American, and a law school graduate who spends his days not lugging lunches but corralling choristers. After escaping practicing law twenty-five years ago—he walked into the office of his firm's senior partner one day and said, "I just can't do this"—this musically inclined son of a Lutheran pastor became a choir director. Now he's the artistic director for the Congressional Chorus in Washington, D.C. And on a frosty Friday night at the end of winter, he's standing in front of eighty singers at the city's Atlas Performing Arts Center as the chorus performs *Road Trip!*—a two-and-a-half-hour show of more than twenty American songs and medleys.

Choirs are peculiar. A lone voice can sing a song. But combine a few voices, sometimes lots of voices, and the result transcends the sum of the parts. Yet bringing all those voices together is challenging, especially for a chorus like this, which is composed entirely of amateurs. The Congressional Chorus earned its name when it began in the mid-1980s as a ragtag group of twelve Capitol Hill staffers seeking a platform for their love of music and an outlet for their frustrations with politics. Today, about one hundred adults—some congressional aides still, but also plenty of lawyers, lobbyists, accountants, marketers, and teachers—perform in the choir. (Washington, D.C., in fact, has more choruses per capita than any city in the U.S.) Many singers have experience in university or religious choirs. Some have genuine talent. But none of them are professionals. And because all of them have other work obligations, they can rehearse only a few times per week.

So how does Simmons keep them in synch? How, during the

evening's California surfer medley, does he get six dozen amateur singers swaying on risers and a half dozen amateur dancers performing in front of them to switch seamlessly—in real time and in front of an audience—from "Surfer Girl" to "I Get Around" and conclude with everyone singing the final sound of the final syllable of the final word of "Surfin' U.S.A." at precisely the same moment?

"I'm a dictator," he tells me. "I work them really hard."

Simmons auditions each member, and he alone decides who's in and who's out. He begins rehearsals precisely at 7 p.m. with each minute mapped out in advance. He selects every piece of music for every concert. (To be more democratic and let members choose what to sing, he says, would turn a concert into a "potluck dinner" rather than a three-star Michelin meal.) He brooks little dissent from the singers. But the reason isn't some deep-seated authoritarian impulse. It's because he's discovered that efficiency in this realm demands firm direction and, occasionally, gentle despotism. As one of his choristers who initially bridled at such leadership once told him, "I always find it amazing that it starts off with nobody knowing anything at the first rehearsal. And by the last concert, you can flick your wrist and we all put the T in the same place."

The first principle of synching fast and slow is that group timing requires a boss—someone or something above and apart from the group itself to set the pace, maintain the standards, and focus the collective mind.

In the early 1990s, a young professor at MIT's Sloan School of Management was frustrated by a gap in the scholarly understanding of how organizations functioned. "Time is arguably the most pervasive aspect of our lives," Deborah Ancona wrote, yet it "has not played a significant nor explicit role in organizational behavioral research." So in a 1992 paper titled "Timing Is Everything," she borrowed a concept from the chronobiology of individuals and applied it to the anthropology of teams.[3]

You'll remember from chapter 1 that within our body and brain are biological clocks that affect our performance, mood, and wakefulness. But you might not recall that those clocks typically run a bit longer than twenty-four hours. Left on our own—say, by spending months in an underground chamber not exposed to light or other people, as in some experiments—our behavior gradually drifts so that before long we're asleep in the afternoons and wide-awake at night.[4] What prevents such misalignment in the aboveground world are environmental and social signals such as sunrise and alarm clocks. The process by which our internal clocks synch up with external cues so we wake up in time for work or go to sleep at a reasonable hour is called "entrainment."

Ancona argued that entrainment also occurs in organizations.[5] Certain activities—product development or marketing—establish their own tempos. But those rhythms necessarily must synchronize with the external rhythms of organizational life—fiscal years, sales cycles, even the age of the company or the stage of people's careers. Just as individuals entrain to external cues, Ancona argued, so do organizations.

In chronobiology, those external cues are known as "zeitgebers" (German for "time giver")—"environmental signals that can synchronize the circadian clock," as Till Roenneberg puts it.[6] Ancona's thinking helped establish that groups also need zeitgebers. Sometimes that pacesetter is a single leader, someone like David Simmons. Indeed, the evidence shows that groups generally attune to the pacing preferences of their highest-status members.[7] However, status and stature are not always one.

Competitive rowing is one of the only racing sports where the athletes have their backs to the finish line. Only one teammate faces forward. And on George Washington University's NCAA Division I women's team, that person was Lydia Barber, the coxswain. In practices and competitions, Barber, who graduated in 2017, sat in the stern of the boat, a headset microphone strapped to her head, shouting instructions at eight rowers. Traditionally, coxswains are as small and light as possible so the boat has less weight to carry. Barber is just four feet tall (she has dwarfism). But her temperament and skills are such a ferocious combination of focus and leadership that, in many ways, she carries the boat.

Barber was the pacesetter, and therefore the boss, for a team of rowers whose 2,000-meter competitions typically last seven minutes. During those 400 to 500 seconds, she called out the rhythm of the strokes, which meant "you must be willing to be in charge and have a big personality," she told me. A race typically begins with the boat sitting in the water, so the rowers must make five quick short strokes just to get moving. Barber next would call out fifteen "high strokes"—at a pace of about forty strokes per minute. Then she'd execute a shift to a slightly slower stroke rhythm, warning her rowers "Shifting one . . . shifting two . . . shifffffft!"

For the rest of the race, her job was to steer the shell, execute the race strategy, and, most important, keep the team motivated and synchronized. In a competition against Duquesne University, this is part of what her call sounded like:

We're RAAAAACCCCIIIING this!
It's BEAUtiful.
Put the blade innnnn . . . and GO!
(beat)
That's one.
(beat)
Two . . .

Load it up!
Three . . .
TAKE that gap!
Four . . .
TAKE that gap!
Five . . .
Run away with it.
Six . . .
Go!
Seven . . .
GO!
Eight . . .
Big LEGGGGS!
Nine . . .
Hell yeah!
Ten . . .
Sit up! Blades in!
Fuck yeah, G-Dubs! Get the legs in and GO!

The boat can't move at its fastest pace without the eight rowers exquisitely synchronized with one another. But they can't synch effectively without Barber. Their speed depends on someone who never touches an oar, just as the Congressional Chorus's sound hinges on Simmons, who never sings a note. For group timing, the boss is above, apart, and essential.

In the case of the dabbawalas, however, the boss—their zeitgeber—doesn't settle in front of a music stand or crouch in the stern of a boat. It hovers above their heads in the train station and in their minds throughout the day.

Most of Ahilu Adhav's morning pickups are quick and efficient—an arm extended from inside an apartment thrusting a bag into Adhav's waiting hands. He doesn't phone ahead of time. Customers don't track him as if he were an Uber or a Lyft car. By the end of his route, he has fifteen bags dangling from his bicycle. He pedals to a patch of pavement across from the Vile Parle train station, where he's soon joined by about ten other men. They unfasten the lunches, pile them on the ground, and start sorting the bags with the speed and self-assurance of a three-card monte dealer. Each person then assembles ten to twenty lunches, ties them together, and slings the bundle over his back. Then they march toward the train station to the platform of the Western line of the Mumbai rail system.

Dabbawalas have considerable autonomy in their jobs. Nobody tells them in what order they must collect or deliver the lunches. They determine the division of labor among the team without anyone acting as a heavy-handed foreman.

But in one dimension, they have no leeway at all: time. Indian business culture typically schedules lunch between 1 p.m. and 2 p.m. That means the dabbawalas must make all their deliveries by 12:45 p.m. And that means Adhav's team must board the 10:51 a.m. train from the Vile Parle station. Miss that train and the entire schedule crumbles. For the dabbawalas, the railway schedule is the boss—the external standard that sets the rhythm, pace, and tempo of their work, the force that imposes discipline on what could otherwise be chaos. It is the unassailable despot, the czarist zeitgeber whose authority is unquestioned and whose rulings are final—an inanimate coxswain or chorus master.

So on this Monday, as on all days, the dabbawalas arrive on the platform with several minutes to spare. As the overhead clock approaches 10:45, they all gather their bags, and before the train has even fully stopped, they clamber into its luggage compartment to ride into South Mumbai.

THE BENEFITS OF BELONGING:
SYNCHING TO THE TRIBE

Here's something you should know about Mumbai's dabbawalas: Most of them have, at best, an eighth grade education. Many of them cannot read or write, a fact that only deepens the implausibility of what they do.

Suppose you're a venture capitalist and I pitch you the following business idea:

It's a lunch-delivery service. Homemade meals picked up at people's apartments and delivered precisely at lunchtime to the desk of their family member on the other side of town. That town, by the way, is the world's tenth largest city, with twice the population of New York City but lacking much of its basic infrastructure. Our venture will not use mobile phones, text messages, online maps, or pretty much any other communications technology. And to staff the operation, we will hire people who have not graduated from secondary school, many of whom are functionally illiterate.

I'm guessing you wouldn't offer me a second meeting, let alone any funding.

Yet Raghunath Medge, president of the Nutan Mumbai Tiffin Box Suppliers Association, claims the dabbawalas have an error rate of 1 in 16 million, a statistic widely repeated but never verified. Still, their efficiency is notable enough to have been celebrated by Richard Branson and Prince Charles—and to have been memorialized in a Harvard Business School case study. Somehow, since its beginnings in 1890, it has worked. And one reason it works is the second principle of group timing.

After individuals synch to the boss, the external standard that sets the pace of their work, they must synch to the tribe—to one another. That requires a deep sense of belonging.

In 1995, two social psychologists, Roy Baumeister and Mark Leary, put forth what they called "the belongingness hypothesis." They proposed that "a need to belong is a fundamental human motivation . . . and that much of what human beings do is done in the service of belongingness." Other thinkers, including Sigmund Freud and Abraham Maslow, had made similar claims, but Baumeister and Leary set about finding empirical proof. The evidence they assembled was overwhelming (their twenty-six-page paper cites more than three hundred sources). Belongingness, they found, profoundly shapes our thoughts and emotions. Its absence leads to ill effects, its presence to health and satisfaction.[8]

Evolution offers at least a partial explanation.[9] After we primates climbed down from trees to roam the open savannah, belonging to a group became essential for survival. We needed others to share the work and watch our backs. Belonging kept us alive. Not belonging turned us into lunch for some prehistoric beast.

Today, this enduring preference for belonging helps us time our actions with others. Social cohesion, many scholars have discovered, leads to greater synchrony.[10] Or, as Simmons puts it, "You get a better sound if there's a sense of belonging. You get better attendance rates at rehearsals, better smiles on their faces." But while the drive for belonging is innate, its emergence sometimes requires some effort. For group coordination, it comes in three forms: codes, garb, and touch.

Codes

For the dabbawalas, the secret code is painted (or written with a marker) on every lunch bag they handle. For example, look at this photograph, taken from a bird's-eye view, of the top of a lunch container that Adhav was transporting:

To you, me, and even the owner of the lunch bag, what's scrawled there is meaningless. But to the dabbawalas, it's the key to coordinating. As our train rumbles toward South Mumbai, and our bodies rumble along with it (this isn't luxury travel), Adhav explains the symbols. VP and Y indicate the neighborhood and building from which the lunch was picked up that morning. The 0 is the station where the lunch will exit. The 7 tells which person will take the lunch from the station to the customer. And the S137 indicates the building and floor where that customer works. That's it. No bar codes, not even any street addresses. "I look at this," Adhav tells me, "and it's all in my head."

In the luggage compartment—nobody's allowed to carry big packages in Mumbai's overstuffed railway cars—the dabbawalas sit on the floor amid a heap of maybe two hundred cloth and plastic lunch bags. They joke and talk with one another in Marathi, the language of the state of Maharashtra, rather than in the far more dominant language of Hindi. The dabbawalas all come from the same set of small villages roughly 150 kilometers southeast of Mumbai. Many are related. Adhav and Medge, in fact, are cousins.

Swapnil Bache, one of the dabbawalas, tells me that the shared language and home villages create what he calls "a brotherly feeling." And that sense of affiliation, like the codes on the lunches, produces an informal understanding that allows them to anticipate one another's actions and move in harmony.

Feelings of belonging boost job satisfaction and performance. Research by Alex Pentland at MIT "has shown that the more cohesive and communicative a team is—the more they chat and gossip—the more they get done."[11] Even the structure of the operation fosters belongingness. The dabbawalas are not a corporation but a cooperative, which operates on a profit-sharing model that pays each worker in equal shares.* Shared language and heritage make it easy to share profits.

Garb

Adhav is thin and wiry. His white shirt fits him more as if his body were a hanger than a mannequin. He wears dark trousers and sandals, and has two bindi dots on his forehead. But atop his head is the most important element of his attire—a white Gandhi hat that

---

* A dabbawala typically earns an average of about $210 per month—not a princely sum by Indian standards but about enough to support a rural family.

signifies that he is a dabbawala. One of the few restrictions on their behavior is that they must wear this hat on the job at all times. The hat is another element of their synchronization. It affiliates them with one another and identifies them to those outside the dabbawala tribe.

Dabbawalas Eknath Khanbar (left) *and Swapnil Bache examine the code that determines where to deliver a lunch.*

Clothing, operating as a marker of affiliation and identification, enables coordination. Take elite restaurants, whose inner workings are one part ballet, another part military invasion. Auguste Escoffier, one of the pioneers of French cuisine, believed that clothing created synchrony. "Escoffier disciplined, drilled, and dressed his chefs," one analyst writes. "Uniforms enforced erect posture and bearing. The double breasted white jacket became the standard to emphasize cleanliness and good sanitation. More subtly, these jackets helped infuse a sense of loyalty, inclusion and pride amongst the chefs, between them and the rest of the restaurant staff."[12]

What's true for French lunch makers is equally true for Indian lunch deliverers.

Touch

Some choirs extend their synchronization to their fingertips. When they sing, they hold hands—to connect to one another and improve the quality of their sound. The dabbawalas don't hold hands. But they do show the physical ease of people who know one another well. They drape an arm around a colleague or pat him on the back. They can communicate with those beyond hearing distance by pointing and using other gestures. And on train rides, in a luggage compartment that lacks discrete seats, they often lean against one another, one man napping on another's shoulder.

Touch is another bolster for belongingness. For example, a few years ago University of California-Berkeley researchers tried to predict the success of NBA basketball teams by examining their use of this tactile language. They watched every team play an early-season game and counted how often the players touched one another—a list that included "fist bumps, high fives, chest bumps, leaping shoulder bumps, chest punches, head slaps, head grabs, low fives, high tens, full hugs, half hugs, and team huddles." Then they monitored team performance over the rest of the season.

Even after controlling for the obvious factors that affect basketball outcomes—for example, the quality of players—they found that touch predicted both individual and team performance. "Touch is the most highly developed sense at birth, and preceded language in hominid evolution," they write. "[T]ouch increases cooperative behavior within groups, which in turn enables better group performance." Touching is a form of synching, a primal way to indicate where you are and where you're going. "Basketball has evolved its own language of touch," they write. "High fives and fist bumps, seemingly small dramatic demonstrations during group interactions, have a lot to say about the cooperative workings of a team, and whether the team wins or loses."[13]

Group timing requires belongingness, which is enabled by codes, garb, and touch. Once groups synch to the tribe, they're ready to synch at the next, and final, level.

EFFORT AND ECSTASY: SYNCHING TO THE HEART

Intermission has ended. The Congressional Chorus singers climb the four risers for act two of *Road Trip!* For the next seventy minutes, they'll sing another dozen songs, including a gorgeous twenty-four-person a cappella rendition of "Baby, What a Big Surprise."

The choristers' voices are in synch, of course. Anyone can hear that. But what's going on inside their bodies, though not audible, is important and intriguing. During this performance, the hearts of this diverse set of amateur singers are likely beating at the same pace.[14]

Synching to the heart is the third principle of group timing. Synchronizing makes us feel good—and feeling good helps a group's wheels turn more smoothly. Coordinating with others also makes us do good—and doing good enhances synchronization.

Exercise is one of the few activities in life that is indisputably good for us—an undertaking that extends enormous benefits but extracts few costs. Exercise helps us live longer. It fends off heart disease and diabetes. It reduces our weight and improves our strength. And its psychological value is enormous. For people suffering from depression, it can be just as effective as medication. For healthy people, it's an instant and long-lasting mood booster.[15] Anyone who examines the science on exercise reaches the same conclusion: People would be silly *not* to do it.

Choral singing might be the new exercise.

The research on the benefits of singing in groups is stunning. Choral singing calms heart rates and boosts endorphin levels.[16] It improves lung function.[17] It increases pain thresholds and reduces the need for pain medication.[18] It even alleviates symptoms of irritable bowel syndrome.[19] Group singing—not just performances but also practices—increases the production of immunoglobulin, making it easier to fight infections.[20] In fact, cancer patients who sing in choirs show an improved immune response after just one rehearsal.[21]

And while the physiological payoffs are many, the psychological ones might be even greater. Several studies demonstrate that choral singing delivers a significant boost to positive mood.[22] It also lifts self-esteem while reducing feelings of stress and symptoms of depression.[23] It enhances one's sense of purpose and meaning, and increases sensitivity toward others.[24] And these effects come not from singing per se but from singing *in a group*. For example, people who sing in choirs report far higher well-being than those who sing solo.[25]

The consequence is a virtuous circle of good feeling and improved coordination. Feeling good promotes social cohesion, which makes it easier to synchronize. Synchronizing with others feels good, which deepens attachment and improves synchronization further still.

Choral groups are the most robust expression of this phenomenon, but other activities in which participants find a way to operate in synch also create similar good feelings. Researchers at the University of Oxford have found that group dancing—"a ubiquitous human activity that involves exertive synchronized movement to music"— raises the pain threshold of people who participate.[26] The same is true for rowing, an endeavor lathered in agony. Other Oxford research, conducted on members of the university's crew team, found elevated pain thresholds when people rowed together but less elevated ones when individuals rowed alone. They even call this state of mind, in which synchronized participants become less susceptible to pain, "rowers' high."[27]

The book *The Boys in the Boat* by Daniel James Brown, which tells the story of a nine-person crew team from the University of Washington that won a gold medal at the 1936 Berlin Olympics, offers an especially vivid description:

> And he came to understand how those almost mystical bonds of trust and affection, if nurtured correctly, might lift a crew above the ordinary sphere, transport it to a place where nine boys somehow became one thing—a thing that could not quite be defined, a thing that was so in tune with the water and the earth and the sky above that, as they rowed, effort was replaced by ecstasy.[28]

That nine individuals can become one humming unit, and that ecstasy can supplant effort as a consequence of that, suggests some deeply ingrained need to synchronize. Some scholars argue that we have an innate desire to feel in pace with others.[29] One Sunday afternoon, I asked David Simmons a question broader than how the Congressional Chorus singers hit their *T*s at the same time. *Why* do human beings sing in groups? I wondered.

He thought about it a moment and answered, "It makes people feel like they're not alone in the world."

Back at the Congressional Chorus concert, a rousing version of "My Shot" from the musical *Hamilton* brings the audience to its feet. The crowd is now synchronized, too, erupting in rhythmic applause and cheers.

The penultimate number, Simmons announces, is "This Land Is Your Land." But before the singers begin, Simmons tells the audience, "We're going to invite you to join us for the final chorus [of the song]. Just watch for my cue." The music starts, the choristers sing. Then Simmons signals the audience with a thrust of his hand, and ever so slowly, three hundred people—most of whom don't know one another and will likely never all be in the same room again—begin singing together, imperfectly but with gusto, until they reach the final line: "This land was made for you and me."

After a forty-minute ride, Ahilu Adhav exits the train at the Marine Lines station, close to where the southern tip of Mumbai meets the Arabian Sea. He's joined by dabbawalas who've arrived from other parts of the city. Using the codes, they quickly sort the bags again. Then Adhav grabs a bicycle another dabbawala has left at this station and sets off to make his deliveries.

This time, though, he can't ride. The streets are so thronged with vehicles, most of them apparently unfamiliar with the concept of lanes, that pushing his bike between stopped cars, revving scooters, and the occasional cow is faster than pedaling it. His first stop is an electrical-parts store on a teeming market street called Vithaldas Lane, where he places a battered lunch bag on the desk of the shop's propri-

Ahilu Adhav delivers two lunches on a busy market street in Mumbai.

etor. The goal is to deliver all the lunches by 12:45 p.m., so his customers (and the dabbawalas themselves) can eat between 1 p.m. and 2 p.m., and Adhav can retrieve the empties in time to board a 2:48 p.m. return train. Today, Adhav completes his rounds at 12:46 p.m.

The previous afternoon Medge, the association president, had described the dabbawalas' jobs to me as a "sacred mission." He tends to talk about lunch delivery in quasireligious terms. He told me that the two critical pillars of the dabbawala creed are that "work is worship" and that the "customer is god." And this heavenly philosophy has an earthly impact. As Medge explained to Stefan Thomke, who wrote the Harvard Business School case study, "If you treat the dabba as a container, then you might not take it seriously. But if you think this container has medicines that must reach patients who are ill and may die, then the sense of urgency forces commitment."[30]

This higher purpose is the dabbawalas' version of synching to the heart. A common mission helps them coordinate, but it also triggers another virtuous circle. Working in harmony with others, science shows, makes it more likely we'll do good. For instance, research by Bahar Tunçgenç and Emma Cohen of the University of Oxford has found that children who played a rhythmic, synchronized clap-and-tap game were more likely than children who played nonsynchronous games to later help their peers.[31] In similar experiments, children who first played synchronous games were far more likely than others to say that if they were to come back for more activities they would be interested in playing with a child who wasn't in their original group.[32] Even swinging in time with another child on a swing set increased subsequent cooperation and collaborative skills.[33] Operating in synch expands our openness to outsiders and makes us more likely to engage in "pro-social" behavior. In other words, coordinating makes us better people—and being better people makes us better coordinators.

Adhav's final tiffin-retrieval stop is at Jayman Industries, a surgical-supply manufacturer with a cramped two-room office. When Adhav arrives, the business's owner, Hitendra Zaveri, hasn't had time to eat yet. So Adhav waits while Zaveri opens his lunch. It's not a sad desk lunch. It looks good—chapatis, rice, dahl, and vegetables.

Zaveri, who's been using the service for twenty-three years, says he prefers a homemade lunch because the quality is assured and because outside food is "not good for the health." He's happy with what he calls the "time accuracy," too. But something subtler keeps him as a customer. His wife cooks his lunch. She's been doing that for a couple of decades. Even though he has a long commute and a frantic day, this brief midday break keeps him connected to her. The dabbawalas make that happen. Adhav's mission might not be exactly sacred, but it's close. He's delivering food—home-cooked food prepared by one family member for another. And he's not doing this once or even once a month. He's doing it almost every single day.

What Adhav does is fundamentally different from delivering a Domino's pizza. He sees one member of a family early in the morning, then another later in the day. He helps the former nourish the latter and the latter appreciate the former. Adhav is the connective tissue that keeps families together. That pizza delivery guy might be efficient, but his work is not transcendent. Adhav, though, is efficient *because* his work is transcendent.

He synchs first to the boss—that 10:51 a.m. train from the Vile Parle station. He synchs next to the tribe—his fellow white-hatted walas who speak the same language and know the cryptic code. But he ultimately synchs to something more sublime—the heart—by doing difficult, physically demanding work that nourishes people and bonds families.

During one of Adhav's morning stops, on the seventh floor of a building called the Pelican, I met a man who has been using the dabbawalas' services for fifteen years. Like so many others I encountered, he says that he's suffered no missed, late, or errant deliveries.

But he did have one complaint.

In the remarkable journey his lunch takes from his own kitchen to Adhav's bicycle to the first train station to a dabbawala's back to another train station to the thronged streets of Mumbai to his office desk, "sometimes your curry is mixed with your rice."

Time Hacker's Handbook

• CHAPTER 6 •

SEVEN WAYS TO FIND YOUR OWN "SYNCHER'S HIGH"

Coordinating and synchronizing with other people is a powerful way to lift your physical and psychological well-being. If your life doesn't involve such activities now, here are some ways to find your own syncher's high:

1. Sing in a chorus.

Even if you've never been part of a musical group, singing with others will instantly deliver a boost. For choral meetups around the world, go to https://www.meetup.com/topics/choir/.

2. Run together.

Running with others offers a trifecta of benefits: exercising, socializing, and synching—all in one. Find a running group through websites like the Road Runners Club of America, http://www.rrca.org/resources/runners/find-a-running-club.

3. Row crew.

Few activities require such perfect synchrony as team rowing. It's also the complete workout: According to some physiologists, a 2,000-meter race burns as many calories as playing back-to-back full-court basketball games. Find a club at http://archive.usrowing .org/domesticrowing/organizations/findaclub.

4. Dance.

Ballroom and other types of social dancing are all about synchronizing with another person and coordinating movements with music. Find a class near you at https://www.thumbtack.com/k/ ballroom-dance-lessons/near-me/.

5. Join a yoga class.

As if you needed to hear one more reason that yoga is good for you, doing it communally may give you a synching high.

6. Flash mob.

For something more adventurous than social dancing and more boisterous than yoga, consider a flash mob—a lighthearted way for strangers to perform for other strangers. They're usually free. And— surprise—most flash mobs are advertised in advance. More info at http://www.makeuseof.com/tag/5-websites-tells-flash-mob-place -organize/.

7. Cook in tandem.

Cooking, eating, and cleaning up by yourself can be a drag. But doing it together requires synchronization and can deliver uplift (not to mention a decent meal). Find tandem-cooking tips at https://www.acouplecooks.com/menu-for-a-cooking-date-tips-for -cooking-together/.

ASK THESE THREE QUESTIONS, THEN KEEP ASKING THEM

Once a group is operating in synch, members' jobs aren't done. Group coordination doesn't abide by the set-it-and-forget-it logic of the Crock-Pot. It requires frequent stirring and a watchful eye. That means to maintain a well-timed group you should regularly—once a week or at least once a month—ask these three questions:

1. Do we have a clear boss—whether a person or some external standard—who engenders respect, whose role is unambiguous, and to whom everyone can direct their initial focus?
2. Are we fostering a sense of belonging that enriches individual identity, deepens affiliation, and allows everyone to synchronize to the tribe?
3. Are we activating the uplift—feeling good and doing good—that is necessary for a group to succeed?

FOUR IMPROV EXERCISES THAT CAN BOOST YOUR GROUP TIMING SKILLS

Improvisational theater requires not just quick thinking but also great synching. Timing your words and movements with other performers without the aid of a script is far more challenging than it looks to an audience. That's why improv groups practice a variety of timing and coordination exercises. Here, recommended by improv guru Cathy Salit, are four that might work for your team:

1. Mirror, Mirror.

Find a partner and face her. Then slowly move your arms or legs—or raise your eyebrows or change your facial expression. Your partner's job is to mirror what you do—to extend her elbow or arch her eyebrow at the same time and same pace as you. Then switch roles and let her act and you mirror. You can also do this in a larger group. Sit in a circle and mirror whatever you see from anyone else in the circle. "This usually starts subtle and then builds until the entire circle is mirroring itself," Salit says.

2. Mind Meld.

This exercise promotes a more conceptual type of synchronization. Find a partner. You count to three together, then each one of you says a word—any word you want—at the same time. Suppose you say "banana" and your partner says "bicycle." Now you both count to three and utter a word that somehow connects the two previous words. In this case, you both might say "seat." Mind meld! But if the two of you offer different words, which is far more likely—suppose one says "store" and the other "wheel"—then the process repeats, counting to three and saying a word that connects "store" and "wheel." Did you both come up with the same word? (I'm thinking "cart"—how about you?) If not, continue until you both say the same word. It's harder than it sounds, but it really builds your mental coordination muscles.

3. Pass the Clap.

This is a classic improv warm-up exercise. Form a circle. The first person turns to his right and makes eye contact with the second person. Then they both clap at the same time. Next, person number two turns to her right, makes eye contact with person number three, and those two clap in unison. (That is, number two passes the clap to number three.) Then number three continues the process. As the clap passes from person to person, somebody can decide to reverse the direction by "clapping back" instead of turning and passing it on. Then anyone else can reverse direction again. The goal is to focus on synching with just one person, which helps the entire group coordinate and pass around an invisible object. Search "pass the clap" on YouTube to see the exercise in action. And while you await your search results, perhaps think of a name for this technique that elicits fewer snickers.

4. Beastie Boys Rap.

Named for the hip-hop group, this group game requires individuals to establish a structure that helps others act in unison. The first person raps a line that follows a particular structure of stressed and unstressed beats. The Improv Resource Center wiki (https://wiki.improvresourcecenter.com) uses this example: "LIVing at HOME is SUCH a DRAG." The rest of the group then follows with this refrain: "YAH buh-buh-BAH buh-BAH buh-BAH BAH!" Then each subsequent person offers a new line, pausing a bit before the final word so the entire group says it together. To continue this example:

> **Person two:** "I always pack my lunch in the same brown BAG."
> **Group:** "YAH buh-buh-BAH buh-BAH buh-BAH BAH!"
> **Person three:** "I like to take a nap on carpet made of SHAG."
> **Group:** "YAH buh-buh-BAH buh-BAH buh-BAH BAH!"

To be clear: Not everyone will instantly warm to all these exercises, but sometimes you've got to fight for your right to synchronize.

FOUR TECHNIQUES FOR PROMOTING BELONGING IN YOUR GROUP

1. Reply quickly to e-mail.

When I asked Congressional Chorus artistic director David Simmons what strategies he used to promote belonging, his answer surprised me. "You reply to their e-mails," he said. The research backs up Simmons's instincts.

E-mail response time is the single best predictor of whether employees are satisfied with their boss, according to research by Duncan Watts, a Columbia University sociologist who is now a principal researcher for Microsoft Research. The longer it takes for a boss to respond to their e-mails, the less satisfied people are with their leader.[1]

2. Tell stories about struggle.

One way that groups cohere is through storytelling. But the stories your group tells should not only be tales of triumph. Stories of failure and vulnerability also foster a sense of belongingness. For instance, Gregory Walton of Stanford University has found that for individuals who might feel apart from a group—for instance, women in a predominantly male environment or students of color in a largely white university—these types of stories can be powerful.[2] Simply reading an account of another student whose freshman year didn't go perfectly but who eventually found her place boosted subsequent feelings of belongingness.

3. Nurture self-organized group rituals.

Cohesive and coordinated groups all have rituals, which help fuse identity and deepen belongingness. But not all rituals have equal power. The most valuable emerge from the people in the group, instead of being orchestrated or imposed by those at the top. For rowers, maybe it's a song they all sing during warm-ups. For choir members, maybe's it's a coffee shop where everyone gathers before each rehearsal. As Stanford's Robb Willer has discovered, "Workplace social functions are less effective if initiated by the manager. What's better are worker-established engagements set at times and places that are convenient for the team."[3] Organic rituals, not artificial ones, generate cohesion.

4. Try a jigsaw classroom.

In the early 1970s, social psychologist Elliot Aronson and his graduate students at the University of Texas designed a cooperative learning technique to address racial divisions in the recently integrated Austin public schools. They called it a "jigsaw classroom." And as it slowly took hold in schools, educators realized the technique could promote group coordination of any kind.

Here's how it works.

The teacher divides students into five-person "jigsaw groups." Then the teacher divides that day's lesson into five segments. For instance, if the class is studying the life of Abraham Lincoln, those sections might be Lincoln's childhood, his early political career, his becoming president at the dawn of the U.S. Civil War, his signing of the Emancipation Proclamation, and his assassination. Each student is responsible for researching one of these segments.

The students then go off to study their piece, forming "expert groups" with students from the class's other five-person groups who share the same assignment. (In other words, all students assigned the Emancipation Proclamation segment meet.) When the research is complete, each student returns to his original jigsaw group and teaches the other four classmates.

The key to this learning strategy is structured interdependence. Each student provides a necessary piece of the whole, something essential for everyone else to glimpse the full picture. And each student's success depends on both her own contribution and those of her partners. If you're a teacher, give it a try. But even if your classroom days are far behind you, you can adapt the jigsaw approach to many work environments.

7.

THINKING IN TENSES

A Few Final Words

Time flies like an arrow. Fruit flies like a banana.

—GROUCHO MARX *(maybe)*

The wisecrack that opens this chapter makes me laugh every time. It's classic Groucho, a language-twisting, brain-bending quip in the tradition of "Outside of a dog, a book is a man's best friend. Inside of a dog, it's too dark to read."[1] Unfortunately, Julius Henry Marx, who became the most famous Marx brother, probably never said it. But the true history of the remark, and the surprisingly complex thought it embodies, offers one final idea for this book.

The real father of these lines, or at least the person who provided the original genetic material, was a linguist, mathematician, and computer scientist named Anthony Oettinger. Today, artificial intelligence and machine learning are white-hot topics, the sources of public fascination and billions of dollars in research and investment. But in the 1950s, when Oettinger began teaching at Harvard University, they were barely known. Oettinger was one of the pioneers in these fields—a multilingual polymath and one of the first people in the world to explore ways that computers could understand natural human language. The quest was, and still is, a challenge.

"Early claims that computers could translate languages were vastly exaggerated," Oettinger wrote in a 1966 *Scientific American* article that predicted with eerie accuracy many of the later scientific uses of computers.[2] The initial difficulty is that many phrases can have multiple meanings when they're removed from a real-life context. The example he used was "Time flies like an arrow." The sentence might mean that time moves with the swiftness of an arrow swooping through the sky. But as Oettinger explained, "time" could also be an imperative verb—a stern instruction to an insect-speed researcher "to take out his stopwatch and time flies with great dispatch, or like an arrow." Or it could be describing a certain species of flying bug—time flies—that exhibit a fondness for arrows. He said programmers could get computers to try to understand the differences among these three meanings, but the underlying set of rules would create a new batch of problems. Those rules couldn't account for syntactically similar but semantically different sentences such as—wait for it—"Fruit flies like a banana." It was a conundrum.

Before long, the sentence "Time flies like an arrow" became a go-to example at conferences and in lectures to illustrate the challenges of machine learning. "The word 'time' here may be either a noun, an adjective, or a verb, yielding three different syntactical interpretations," wrote Frederick Crosson, a University of Notre Dame professor and editor of one of the first artificial intelligence textbooks.[3] The arrow-banana pairing endured and, years later, somehow became attached to Groucho Marx. But Yale librarian and quotation guru Fred Shapiro says, "There is no reason to believe that Groucho actually said this."[4]

Yet the sturdiness of the line reveals something important. As Crosson points out, even in a five-word sentence, "time" can function as a noun, an adjective, or a verb. It is one of the most expansive and versatile words we have. "Time" can be a proper noun, as in "Greenwich Mean Time." The noun form can also signify a discrete duration ("How much time is left in the second period?"), a specific moment ("What time does the bus to Narita arrive?"), an abstract notion ("Where did the time go?"), a general experience ("I'm having a good time"), a turn at doing something ("He rode the roller coaster only one time"), a historical period ("In Winston Churchill's time . . ."), and more. In fact, according to Oxford University Press researchers, "time" is the most common noun in the English language.[5]

As a verb, it also has multiple meanings. We can time a race, which always involves a clock, or time an attack, which often does not. We can time, as in keeping time, when playing a musical instrument. And we, like dabbawalas and rowers, can time our actions with others. The word can function as an adjective, as in "time bomb," "time zone," and "time clock"—and "adverbs of time" represent an entire category of that part of speech.

But time pervades our language and colors our thought even more deeply. Most of the world's languages mark verbs with time using tenses—especially past, present, and future—to convey meaning and reveal thinking. Nearly every phrase we utter is tinged with time. In some sense, we think in tenses. And that's especially true when we think about ourselves.

Consider the past. It's something we're told not to dwell on, but research makes it clear that thinking in the past tense can lead to a greater understanding of ourselves. For instance, nostalgia—contemplating and sometimes aching for the past—was once considered a pathology, an impairment that diverted us from current goals. Scholars of the seventeenth and eighteenth centuries thought it was a physical ailment—"a cerebral disease of essentially demonic cause" spurred by "the quite continuous vibration of animal spirits through

[the] fibers of the middle brain." Others believed nostalgia was caused by changes in atmospheric pressure or "an oversupply of black bile in the blood" or was perhaps an affliction unique to the Swiss. By the ninteenth century, those ideas were discarded, but the pathologizing of nostalgia was not. Scholars and physicians of that era believed it was a mental dysfunction, a psychiatric disorder connected to psychosis, compulsion, and Oedipal yearnings.[6]

Today, thanks to the work of psychologist Constantine Sedikides of the University of Southampton and others, nostalgia has been redeemed. Sedikides calls it "a vital intrapersonal resource that contributes to psychological equanimity . . . a repository of psychological sustenance." The benefits of thinking fondly about the past are vast because nostalgia delivers two ingredients essential to well-being: a sense of meaning and a connection to others. When we think nostalgically, we often feature ourselves as the protagonist in a momentous event (a wedding or a graduation, for instance) that involves the people we care about most deeply.[7] Nostalgia, research shows, can foster positive mood, protect against anxiety and stress, and boost creativity.[8] It can heighten optimism, deepen empathy, and alleviate boredom.[9] Nostalgia can even increase *physiological* feelings of comfort and warmth. We're more likely to feel nostalgic on chillier days. And when experimenters induce nostalgia—through music or smell, for instance—people are more tolerant of cold and perceive the temperature to be higher.[10]

Like poignancy, nostalgia is a "bittersweet but predominantly positive and fundamentally social emotion." Thinking in the past tense offers "a window into the intrinsic self," a portal to who we really are.[11] It makes the present meaningful.

The same principle applies to the future. Two prominent social scientists—Daniel Gilbert of Harvard University and Timothy Wilson of the University of Virginia—have argued that while "all animals are on a voyage through time," humans have an edge. Antelope and salamanders can predict the consequences of events they've experienced before. But only humans can "pre-experience" the future by simulating it in our minds, what Gilbert and Wilson call "prospection."[12] However, we're not nearly as skilled in this ability as we believe we are. While the reasons vary, the language we speak—literally the tenses we use—can play a role.

M. Keith Chen, an economist now at UCLA, was one of the first to explore the connection between language and economic behavior. He first grouped thirty-six languages into two categories—those that have a strong future tense and those that have a weak or nonexistent one. Chen, an American who grew up in a Chinese-speaking household, offers the differences between English and Mandarin to illustrate the distinction. He says, "[I]f I wanted to explain to an English-speaking colleague why I can't attend a meeting later today, I could not say 'I go to a seminar.'" In English, Chen would have to explicitly mark the future by saying, "I *will be going* to a seminar" or "I *have to* go to a seminar." However, Chen says, if "on the other hand I were speaking Mandarin, it would be quite natural for me to omit any marker of future time and say *Wǒ qù tīng jiǎngzò* (I *go* listen seminar)."[13] Strong-future languages such as English, Italian, and Korean require speakers to make sharp distinctions between the present and the future. Weak-future languages such as Mandarin, Finnish, and Estonian draw little or often no contrast at all.

Chen then examined—controlling for income, education, age, and other factors—whether people speaking strong-future and weak-future languages behaved differently. They do—in somewhat stunning fashion. Chen found that speakers of weak-future languages—those that did not mark explicit differences between present and future—were 30 percent more likely to save for retirement and 24 percent less likely to smoke. They also practiced safer sex, exercised more regularly, and were both healthier and wealthier in retirement. This was true even within countries such as Switzerland, where some citizens spoke a weak-future language (German) and others a strong-future one (French).[14]

Chen didn't conclude that the language a person speaks *caused* this behavior. It could merely *reflect* deeper differences. And the question of whether language actually shapes thought and therefore action remains a contentious issue in the field of linguistics.[15] Nonetheless, other research has shown we plan more effectively and behave more responsibly when the future feels more closely connected to the current moment and our current selves. For example, one reason some people don't save for retirement is that they somehow consider the future version of themselves a different person than the current version. But showing people age-advanced images of their own photographs can boost their propensity to save.[16] Other research has found that simply thinking of the future in smaller time units—days, not years—"made people feel closer to their future self and less likely to feel that their current and future selves were not really the same person."[17] As with nostalgia, the highest function of the future is to enhance the significance of the present.

Which leads to the present itself. Two final studies are illuminating. In the first, five Harvard researchers asked people to make small "time capsules" of the present moment (three songs they recently listened to, an inside joke, the last social event they attended, a recent photo, etc.) or write about a recent conversation. Then they asked people to guess how curious they'd be to see what they documented several months later. When the time came to view the time capsules, people were far more curious than they had predicted. They also found the contents of what they'd memorialized far more meaningful than they had expected. Across multiple experiments, people underestimated the value of rediscovering current experiences in the future.

"By recording ordinary moments today, one can make the present a 'present' for the future," the researchers write.[18]

The other study examined the effect of awe. Awe lives "in the upper reaches of pleasure and on the boundary of fear," as two scholars put it. It "is a little studied emotion . . . central to the experience of religion, politics, nature, and art."[19] It has two key attributes: vastness (the experience of something larger than ourselves) and accommodation (the vastness forces us to adjust our mental structures).

Melanie Rudd, Kathleen Vohs, and Jennifer Aaker found that the experience of awe—the sight of the Grand Canyon, the birth of a child, a spectacular thunderstorm—changes our perception of time. When we experience awe, time slows down. It expands. We feel like we have more of it. And that sensation lifts our well-being. "Experiences of awe bring people into the present moment, and being in the present moment underlies awe's capacity to adjust time perception, influence decisions, and make life feel more satisfying than it would otherwise."[20]

Taken together, all of these studies suggest that the path to a life of meaning and significance isn't to "live in the present" as so many spiritual gurus have advised. It is to integrate our perspectives on time into a coherent whole, one that helps us comprehend who we are and why we're here.

In an otherwise forgettable scene in the 1930 movie *Animal Crackers*, Groucho Marx corrects himself for using the verb "are" when he should have said "were." He explains, "I was using the subjunctive instead of the past tense." Then, after a beat, he adds, "We're way past tents, we're living in bungalows now."

We, too, are way beyond tenses. The challenge of the human condition is to bring the past, present, and future together.

When I began working on this book, I knew that timing was important, but also that it was inscrutable. At the start of this project, I had no idea of the destination. My goal was to arrive at something resembling the truth, to pin down facts and insights that could help people, including me, work a little smarter and live a little better.

The *product* of writing—this book—contains more answers than questions. But the *process* of writing is the opposite. Writing is an act of discovering what you think and what you believe.

I used to believe in ignoring the waves of the day. Now I believe in surfing them.

I used to believe that lunch breaks, naps, and taking walks were niceties. Now I believe they're necessities.

I used to believe that the best way to overcome a bad start at work, at school, or at home was to shake it off and move on. Now I believe the better approach is to start again or start together.

I used to believe that midpoints didn't matter—mostly because I was oblivious to their very existence. Now I believe that midpoints illustrate something fundamental about how people behave and how the world works.

I used to believe in the value of happy endings. Now I believe that the power of endings rests not in their unmitigated sunniness but in their poignancy and meaning.

I used to believe that synchronizing with others was merely a mechanical process. Now I believe that it requires a sense of belonging, rewards a sense of purpose, and reveals a part of our nature.

I used to believe that timing was everything. Now I believe that everything is timing.

FURTHER READING

Time and timing are endlessly interesting topics that other authors have explored with skill and gusto. Here are seven books, listed in alphabetical order by title, that will deepen your understanding:

168 Hours: You Have More Time Than You Think (2010)
By Laura Vanderkam
We each get the same allotment: 168 hours each week. Vanderkam offers shrewd, actionable advice on how to make the most of those hours by setting priorities, eliminating nonessentials, and focusing on what truly matters.

A Geography of Time: Temporal Misadventures of
  a Social Psychologist (1997)
By Robert V. Levine
Why do some cultures move fast and others slowly? Why do some abide by strict "clock time" and others by more fluid "event time"? A behavioral scientist offers some fascinating answers, many based on his own peripatetic adventures.

Daily Rituals: How Artists Work (2013)
Edited by Mason Currey
How have the world's greatest creators organized their time? This book reveals the daily habits of a range of creative powerhouses—Agatha Christie, Sylvia Plath, Charles Darwin, Toni Morrison, Andy Warhol, and 156 others.

Internal Time: Chronotypes, Social Jet Lag, and Why You're So Tired (2012)
By Till Roenneberg
If you're going to read one book about chronobiology, make it this one. You'll learn more from this smart, concise work—organized into twenty-four chapters to represent the twenty-four hours of the day—than from any other single source.

The Circadian Code: Lose Weight, Supercharge Your Energy, and Transform Your Health from Morning to Midnight (2018)
By Satchin Panda
One of the world's leading circadian rhythm scientists looks at intriguing early research on the "when" of eating. It turns out that restricting our food intake to an 8- or 10-hour window each day could help some people lose weight and better manage conditions such as diabetes.

The Dance of Life: The Other Dimension of Time (1983)
By Edward T. Hall
An American anthropologist examines how cultures around the world perceive time. The analysis is occasionally a bit dated, but the insights are powerful, which is why this book remains a staple of college courses.

Why Time Flies: A Mostly Scientific Investigation (2017)
By Alan Burdick
A wonderful and witty work of science journalism that captures the complexity, frustration, and exhilaration of trying to understand the nature of time.

ACKNOWLEDGMENTS

I f you read acknowledgments—and it looks like you do—you've probably noticed a phenomenon similar to Laura Carstensen's discovery about the declining size of social networks as people age. In their first book, authors typically thank a preposterously wide circle of contacts. ("My third grade gym teacher helped me overcome my fear of rope climbing, perhaps the most vital lesson I've learned as a writer.")

But with each subsequent book, the list shortens. The acknowledgments shrink to the inner circle. Here's mine:

Cameron French was as dedicated and productive a researcher as any author could expect. He filled gigabytes of Dropbox folders with research papers and literature reviews, honed many of the tools and tips, and checked every fact and citation. What's more, he did these things with such intelligence, conscientiousness, and good cheer that I'm tempted in the future to work only with people who grew up in Oregon and went to Swarthmore College. Shreyas Raghavan, now a PhD student at Duke University's Fuqua School of Business, located some of the book's best examples, regularly pushed challenging counterarguments, and patiently explained statistical techniques and quantitative analyses that eluded my short grasp.

Rafe Sagalyn, my literary agent and friend of two decades, was his usual spectacular self. At every stage of the process—developing the idea, producing the manuscript, telling the world about the result— he was indispensable.

At Riverhead Books, the sagacious and perspicacious Jake Morrissey read the text multiple times and lavished attention on every page. His stream of comments and questions—"This isn't a television script"; "Is that the right word?"; "You can go deeper here"— were frequently annoying and invariably correct. I'm also lucky to have on my side Jake's fellow publishing all-stars: Katie Freeman, Lydia Hirt, Geoff Kloske, and Kate Stark.

Tanya Maiboroda created nearly two dozen charts that captured key ideas with clarity and grace. Elizabeth McCullough, as always, caught mistakes in the text that everyone else had missed. Rajesh Padmashali was a brilliant partner, fixer, and translator in Mumbai. Jon Auerbach, Marc Tetel, and Renée Zuckerbrot, friends since my freshman year in college, helped identify several interview subjects. I also benefited from conversations with Adam Grant, Chip Heath, and Bob Sutton, all of whom offered smart research suggestions and one of whom (Adam) talked me out of my atrocious initial outline. Special thanks, too, to Francesco Cirillo and the late Amar Bose for reasons they would understand.

When I first began writing books, one of our kids was tiny and two were nonexistent. Today, all three are astonishing young people who are often willing to help their less astonishing father. Sophia Pink read several chapters and offered an array of astute edits. Saul Pink's considerable basketball acumen—coupled with his phone-based research skills—delivered the great sports tale in chapter 4. Eliza Pink, who navigated her senior year in high school while I finished this book, was my role model for grit and dedication.

And at the center is their mother. Jessica Lerner read every word of this book. But that's not all. She also read every word of this book *out loud.* (If you don't grok how heroic that is, open the introduction, begin reading aloud, and see how far you get. Then try it with someone who constantly interrupts because you're not reading with sufficient verve or proper emphasis.) Her brainpower and empathy made this a better book, just as for a quarter century they've made me a better person. For every time and in every tense, she was, is, and will be the love of my life.

NOTES

INTRODUCTION. CAPTAIN TURNER'S DECISION

1. Tad Fitch and Michael Poirier, *Into the Danger Zone: Sea Crossings of the First World War* (Stroud, UK: The History Press, 2014), 108.
2. Erik Larson, *Dead Wake: The Last Crossing of the Lusitania* (New York: Broadway Books, 2016), 1.
3. Colin Simpson, "A Great Liner with Too Many Secrets," *Life,* October 13, 1972, 58.
4. Fitch and Poirier, *Into the Danger Zone,* 118; Adolph A. Hoehling and Mary Hoehling, *The Last Voyage of the Lusitania* (Lanham, MD: Madison Books, 1996), 247.
5. Daniel Joseph Boorstin, *The Discoverers: A History of Man's Search to Know His World and Himself* (New York: Vintage, 1985), 1.

CHAPTER 1. THE HIDDEN PATTERN OF EVERYDAY LIFE

1. Kit Smith, "44 Twitter Statistics for 2016," *Brandwatch,* May 17, 2016, available at https://www.brandwatch.com/2016/05/44-twitter-stats-2016.
2. Scott A. Golder and Michael W. Macy, "Diurnal and Seasonal Mood Vary with Work, Sleep, and Daylength Across Diverse Cultures," *Science* 333, no. 6051 (2011): 1878–81. Please note that this research was conducted before Donald Trump was elected president and his tweets became part of the political conversation.
3. For a fuller account of de Marian's discovery, see Till Roenneberg, *Internal Time: Chronotypes, Social Jet Lag, and Why You're So Tired* (Cambridge, MA: Harvard University Press, 2012), 31–35.
4. William J. Cromie, "Human Biological Clock Set Back an Hour," *Harvard University Gazette,* July 15, 1999.
5. Peter Sheridan Dodds et al., "Temporal Patterns of Happiness and Information in a Global Social Network: Hedonometrics and Twitter," *PloS ONE* 6, no. 12 (2011): e26752. See also Riccardo Fusaroli et al., "Timescales of Massive Human Entrainment," *PloS ONE* 10, no. 4 (2015): e0122742.

6. Daniel Kahneman et al., "A Survey Method for Characterizing Daily Life Experience: The Day Reconstruction Method," *Science* 306, no. 5702 (2004): 1776–80.

7. Arthur A. Stone et al., "A Population Approach to the Study of Emotion: Diurnal Rhythms of a Working Day Examined with the Day Reconstruction Method," *Emotion* 6, no. 1 (2006): 139–49.

8. Jing Chen, Baruch Lev, and Elizabeth Demers, "The Dangers of Late-Afternoon Earnings Calls," *Harvard Business Review*, October 2013.

9. Ibid.

10. Jing Chen, Elizabeth Demers, and Baruch Lev, "Oh What a Beautiful Morning! Diurnal Variations in Executives' and Analysts' Behavior: Evidence from Conference Calls." Found at https://www.darden.virginia.edu.uploadedfiles/darden_web/content /faculty_research/seminars_and_conferences/CDL_March_2016.pdf.

11. Ibid.

12. Amos Tversky and Daniel Kahneman, "Extensional Versus Intuitive Reasoning: The Conjunction Fallacy in Probability Judgment," *Psychological Review* 90, no. 4 (1983): 293–315.

13. Galen V. Bodenhausen, "Stereotypes as Judgmental Heuristics: Evidence of Circadian Variations in Discrimination," *Psychological Science* 1, no. 5 (1990): 319–22.

14. Ibid.

15. Russell G. Foster and Leon Kreitzman, *Rhythms of Life: The Biological Clocks That Control the Daily Lives of Every Living Thing* (New Haven, CT: Yale University Press, 2005), 11.

16. Carolyn B. Hines, "Time-of-Day Effects on Human Performance," *Journal of Catholic Education* 7, no. 3 (2004): 390–413, citing Tamsin L. Kelly, *Circadian Rhythms: Importance for Models of Cognitive Performance*, U.S. Naval Health Research Center Report, no. 96-1 (1996): 1–24.

17. Simon Folkard, "Diurnal Variation in Logical Reasoning," *British Journal of Psychology* 66, no. 1 (1975): 1–8; Timothy H. Monk et al., "Circadian Determinants of Subjective Alertness," *Journal of Biological Rhythms* 4, no. 4 (1989): 393–404.

18. Robert L. Matchock and J. Toby Mordkoff, "Chronotype and Time-of-Day Influences on the Alerting, Orienting, and Executive Components of Attention," *Experimental Brain Research* 192, no. 2 (2009): 189–198.

19. Hans Henrik Sievertsen, Francesca Gino, and Marco Piovesan, "Cognitive Fatigue Influences Students' Performance on Standardized Tests," *Proceedings of the National Academy of Sciences* 113, no. 10 (2016): 2621–24.

20. Nolan G. Pope, "How the Time of Day Affects Productivity: Evidence from School Schedules," *Review of Economics and Statistics* 98, no. 1 (2016): 1–11.

21. Mareike B. Wieth and Rose T. Zacks, "Time of Day Effects on Problem Solving: When the Non-optimal Is Optimal," *Thinking & Reasoning* 17, no. 4 (2011): 387–401.

22. Lynn Hasher, Rose T. Zacks, and Cynthia P. May, "Inhibitory Control, Circadian Arousal, and Age," in Daniel Gopher and Asher Koriat, eds., *Attention and Performance XVII: Cognitive Regulation of Performance: Interaction of Theory and Application* (Cambridge, MA: MIT Press, 1999), 653–75.

23. Cindi May, "The Inspiration Paradox: Your Best Creative Time Is Not When You Think," *Scientific American*, March 6, 2012.

24. Mareike B. Wieth and Rose T. Zacks, "Time of Day Effects on Problem Solving: When the Non-optimal Is Optimal," *Thinking & Reasoning* 17, no. 4 (2011): 387–401.

25. Inez Nellie Canfield McFee, *The Story of Thomas A. Edison* (New York: Barse & Hopkins, 1922).

26. Till Roenneberg et al., "Epidemiology of the Human Circadian Clock," *Sleep Medicine Reviews* 11, no. 6 (2007): 429–38.

27. Ana Adan et al., "Circadian Typology: A Comprehensive Review," *Chronobiology International* 29, no. 9 (2012): 1153–75; Franzis Preckel et al., "Chronotype, Cognitive Abilities, and Academic Achievement: A Meta-Analytic Investigation," *Learning and Individual Differences* 21, no. 5 (2011): 483–92; Till Roenneberg, Anna Wirz-Justice, and Martha Merrow, "Life Between Clocks: Daily Temporal Patterns of Human Chronotypes," *Journal of Biological Rhythms* 18, no. 1 (2003): 80–90; Iwona Chelminski et al., "Horne and Ostberg Questionnaire: A Score Distribution in a Large Sample of Young Adults," *Personality and Individual Differences* 23, no. 4 (1997): 647–52; G. M. Cavallera and S. Giudici, "Morningness and Eveningness Personality: A Survey in Literature from 1995 up till 2006." *Personality and Individual Differences* 44, no. 1 (2008): 3–21.

28. Renuka Rayasam, "Why Sleeping In Could Make You a Better Worker," *BBC Capital*, February 25, 2016.

29. Markku Koskenvuo et al., "Heritability of Diurnal Type: A Nationwide Study of 8753 Adult Twin Pairs," *Journal of Sleep Research* 16, no. 2 (2007): 156–62; Yoon-Mi Hur, Thomas J. Bouchard, Jr., and David T. Lykken, "Genetic and Environmental Influence on Morningness-Eveningness," *Personality and Individual Differences* 25, no. 5 (1998): 917–25.

30. One possible explanation: Those born in seasons with less light reach their daily circadian peak earlier in order to take advantage of the limited light. Vincenzo Natale and Ana Adan, "Season of Birth Modulates Morning-Eveningness Preference in Humans," *Neuroscience Letters* 274, no. 2 (1999): 139–41; Hervé Caci et al., "Transcultural Properties of the Composite Scale of Morningness: The Relevance of the 'Morning Affect' Factor," *Chronobiology International* 22, no. 3 (2005): 523–40.

31. Till Roenneberg et al., "A Marker for the End of Adolescence," *Current Biology* 14, no. 24 (2004): R1038–39.

32. Till Roenneberg et al., "Epidemiology of the Human Circadian Clock," *Sleep Medicine Reviews* 11, no. 6 (2007): 429–38; See also Ana Adan et al., "Circadian Typology: A Comprehensive Review," *Chronobiology International* 29, no. 9 (2012): 1153–75.

33. Ana Adan et al., "Circadian Typology: A Comprehensive Review," *Chronobiology International* 29, no. 9 (2012): 1153–75. See also Ryan J. Walker et al., "Age, the Big Five, and Time-of-Day Preference: A Mediational Model," *Personality and Individual Differences* 56 (2014): 170–74; Christoph Randler, "Proactive People Are Morning People," *Journal of Applied Social Psychology* 39, no. 12 (2009): 2787–97; Hervé Caci, Philippe Robert, and Patrice Boyer, "Novelty Seekers and Impulsive Subjects Are Low in Morningness," *European Psychiatry* 19, no. 2 (2004): 79–84; Maciej Stolarski, Maria Ledzińska, and Gerald Matthews, "Morning Is Tomorrow, Evening Is Today: Relationships Between Chronotype and Time Perspective," *Biological Rhythm Research* 44, no. 2 (2013): 181–96.

34. Renée K. Biss and Lynn Hasher, "Happy as a Lark: Morning-Type Younger and Older Adults Are Higher in Positive Affect," *Emotion* 12, no. 3 (2012): 437–41.

35. Ryan J. Walker et al., "Age, the Big Five, and Time-of-Day Preference: A Mediational Model," *Personality and Individual Differences* 56 (2014): 170–74; Christoph Randler, "Morningness-Eveningness, Sleep-Wake Variables and Big Five Personality Factors," *Personality and Individual Differences* 45, no. 2 (2008): 191–96.

36. Ana Adan, "Chronotype and Personality Factors in the Daily Consumption of Alcohol and Psychostimulants," *Addiction* 89, no. 4 (1994): 455–62.

37. Ji Hee Yu et al., "Evening Chronotype Is Associated with Metabolic Disorders and Body Composition in Middle-Aged Adults," *Journal of Clinical Endocrinology & Metabolism* 100, no. 4 (2015): 1494–1502; Seog Ju Kim et al., "Age as a Moderator of the Association Between Depressive Symptoms and Morningness-Eveningness," *Journal of Psychosomatic Research* 68, no. 2 (2010): 159–164; Iwona Chelminski et al., "Horne and Ostberg Questionnaire: A Score Distribution in a Large Sample of Young Adults," *Personality and Individual Differences* 23, no. 4 (1997): 647–52; Michael D. Drennan et al., "The Effects of Depression and Age on the Horne-Ostberg Morningness-Eveningness Score," *Journal of Affective Disorders* 23, no. 2 (1991): 93–98; Christoph Randler et al., "Eveningness Is Related to Men's Mating Success," *Personality and Individual Differences* 53, no. 3 (2012): 263–67; J. Kasof, "Eveningness and Bulimic Behavior," *Personality and Individual Differences* 31, no. 3 (2001): 361–69.

38. Kai Chi Yam, Ryan Fehr, and Christopher M. Barnes, "Morning Employees Are Perceived as Better Employees: Employees' Start Times Influence Supervisor Performance Ratings," *Journal of Applied Psychology* 99, no. 6 (2014): 1288–99.

39. Catharine Gale and Christopher Martyn, "Larks and Owls and Health, Wealth, and Wisdom," *British Medical Journal* 317, no. 7174 (1998): 1675–77.

40. Richard D. Roberts and Patrick C. Kyllonen, "Morning-Eveningness and Intelligence: Early to Bed, Early to Rise Will Make You Anything but Wise!," *Personality and Individual Differences* 27 (1999): 1123–33; Davide Piffer et al., "Morning-Eveningness and Intelligence Among High-Achieving US Students: Night Owls Have Higher GMAT Scores than Early Morning Types in a Top-Ranked MBA Program," *Intelligence* 47 (2014): 107–12.

41. Christoph Randler, "Evening Types Among German University Students Score Higher on Sense of Humor After Controlling for Big Five Personality Factors," *Psychological Reports* 103, no. 2 (2008): 361–70.

42. Galen V. Bodenhausen, "Stereotypes as Judgmental Heuristics: Evidence of Circadian Variations in Discrimination," *Psychological Science* 1, no. 5 (1990): 319–22.

43. Mareike B. Wieth and Rose T. Zacks, "Time-of-Day Effects on Problem Solving: When the Non-optimal is Optimal," *Thinking & Reasoning* 17, no. 4 (2011): 387–401.

44. Cynthia P. May and Lynn Hasher, "Synchrony Effects in Inhibitory Control over Thought and Action," *Journal of Experimental Psychology: Human Perception and Performance* 24, no. 2 (1998): 363–79; Ana Adan et al., "Circadian Typology: A Comprehensive Review," *Chronobiology International* 29, no. 9 (2012): 1153–75.

45. Ángel Correa, Enrique Molina, and Daniel Sanabria, "Effects of Chronotype and Time of Day on the Vigilance Decrement During Simulated Driving," *Accident Analysis & Prevention* 67 (2014): 113–18.

46. John A. E. Anderson et al., "Timing Is Everything: Age Differences in the Cognitive Control Network Are Modulated by Time of Day," *Psychology and Aging* 29, no. 3 (2014): 648–58.

47. Brian C. Gunia, Christopher M. Barnes, and Sunita Sah, "The Morality of Larks and Owls: Unethical Behavior Depends on Chronotype as Well as Time of Day," *Psychological Science* 25, no. 12 (2014): 2272–74; Maryam Kouchaki and Isaac H. Smith, "The Morning Morality Effect; The Influence of Time of Day on Unethical Behavior," *Psychological Science* 25, no. 1 (2013): 95–102.

48. Mason Currey, ed., *Daily Rituals: How Artists Work* (New York: Knopf, 2013), 62–63.

49. Ibid., 29–32, 62–63.

50. Céline Vetter et al., "Aligning Work and Circadian Time in Shift Workers Improves Sleep and Reduces Circadian Disruption," *Current Biology* 25, no. 7 (2015): 907–11.

CHAPTER 1. TIME HACKER'S HANDBOOK

1. Karen Van Proeyen et al., "Training in the Fasted State Improves Glucose Tolerance During Fat-Rich Diet," *Journal of Physiology* 588, no. 21 (2010): 4289–302.

2. Michael R. Deschenes et al., "Chronobiological Effects on Exercise: Performance and Selected Physiological Responses," *European Journal of Applied Physiology and Occupational Physiology* 77, no. 3 (1998): 249–560.

3. Elise Facer-Childs and Roland Brandstaetter, "The Impact of Circadian Phenotype and Time Since Awakening on Diurnal Performance in Athletes," *Current Biology* 25, no. 4 (2015): 518–22.

4. Boris I. Medarov, Valentin A. Pavlov, and Leonard Rossoff, "Diurnal Variations in Human Pulmonary Function," *International Journal of Clinical Experimental Medicine* 1, no. 3 (2008): 267–73.

5. Barry Drust et al., "Circadian Rhythms in Sports Performance: An Update," *Chronobiology International* 22, no. 1 (2005), 21–44; João Paulo P. Rosa et al., "2016 Rio Olympic Games: Can the Schedule of Events Compromise Athletes' Performance?" *Chronobiology International* 33, no. 4 (2016): 435–40.

6. American Council on Exercise, "The Best Time to Exercise," *Fit Facts* (2013), available at https://www.acefitness.org/fitfacts/pdfs/fitfacts/itemid_2625.pdf.

7. Miguel Debono et al., "Modified-Release Hydrocortisone to Provide Circadian Cortisol Profiles," *Journal of Clinical Endocrinology & Metabolism* 94, no. 5 (2009): 1548–54.

8. Alicia E. Meuret et al., "Timing Matters: Endogenous Cortisol Mediates Benefits from Early-Day Psychotherapy," *Psychoneuroendocrinology* 74 (2016): 197–202.

CHAPTER 2. AFTERNOONS AND COFFEE SPOONS

1. Melanie Clay Wright et al., "Time of Day Effects on the Incidence of Anesthetic Adverse Events," *Quality and Safety in Health Care* 15, no. 4 (2006): 258–63. The quotation at the end of the paragraph is from the lead author on the paper and appears in "Time of Surgery Influences Rate of Adverse Health Events Due to Anesthesia," *Duke News*, August 3, 2006.

2. Alexander Lee et al., "Queue Position in the Endoscopic Schedule Impacts Effectiveness of Colonoscopy," *American Journal of Gastroenterology* 106, no. 8 (2011): 1457–65.

3. One study found a gender difference, concluding that "[c]olonoscopies performed in the afternoon tend to have lower polyp and adenoma detection rates . . . [but] the lower adenoma detection rate in afternoon colonoscopies seems to apply mainly to female patients." Shailendra Singh et al., "Differences Between Morning and Afternoon Colonoscopies for Adenoma Detection in Female and Male Patients," *Annals of Gastroenterology* 29, no. 4 (2016): 497–501. A few other studies have been more circumspect about the time-of-day effect. See, e.g., Jerome D. Waye, "Should All Colonoscopies Be Performed in the Morning?" *Nature Reviews: Gastroenterology & Hepatology* 4, no. 7 (2007): 366–67.

4. Madhusudhan R. Sanaka et al., "Afternoon Colonoscopies Have Higher Failure Rates Than Morning Colonoscopies," *American Journal of Gastroenterology* 101, no. 12 (2006): 2726–30; Jerome D. Waye, "Should All Colonoscopies Be Performed in the Morning?" *Nature Reviews: Gastroenterology & Hepatology* 4, no. 7 (2007): 366–67.

5. Jeffrey A. Linder et al., "Time of Day and the Decision to Prescribe Antibiotics," *JAMA Internal Medicine* 174, no. 12 (2014): 2029–31.

6. Hengchen Dai et al., "The Impact of Time at Work and Time Off from Work on Rule Compliance: The Case of Hand Hygiene in Health Care," *Journal of Applied Psychology* 100, no. 3 (2015): 846–62. the 38 percent figure represents "the fitted odds of compliance over the course of a 12-hr shift or an 8.7-percentage-point decrease in the rate of compliance for an average caregiver over the course of a 12-hr shift."

7. Ibid.

8. Jim Horne and Louise Reyner, "Vehicle Accidents Related to Sleep: A Review," *Occupational and Environmental Medicine* 56, no. 5 (1999): 289–94.

9. Justin Caba, "Least Productive Time of the Day Officially Determined to Be 2:55 PM: What You Can Do to Stay Awake?" *Medical Daily,* June 4, 2013, available at http://www.medicaldaily.com/least-productive-time-day-officially-determined-be -255-pm-what-you-can-do-stay-awake-246495.

10. Maryam Kouchaki and Isaac H. Smith, "The Morning Morality Effect: The Influence of Time of Day on Unethical Behavior," *Psychological Science* 25, no. 1 (2014): 95–102; Maryam Kouchaki, "In the Afternoon, the Moral Slope Gets Slipperier," *Harvard Business Review,* May 2014.

11. Julia Neily et al., "Association Between Implementation of a Medical Team Training Program and Surgical Mortality," *JAMA* 304, no. 15 (2010): 1693–1700.

12. Hans Henrik Sievertsen, Francesca Gino, and Marco Piovesan, "Cognitive Fatigue Influences Students' Performance on Standardized Tests," *Proceedings of the National Academy of Sciences* 113, no. 10 (2016): 2621–24.

13. Francesca Gino, "Don't Make Important Decisions Late in the Day," *Harvard Business Review,* February 23, 2016.

14. Hans Henrik Sievertsen, Francesca Gino, and Marco Piovesan, "Cognitive Fatigue Influences Students' Performance on Standardized Tests," *Proceedings of the National Academy of Sciences* 113, no. 10 (2016): 2621–24.

15. Kyoungmin Cho, Christopher M. Barnes, and Cristiano L. Guanara, "Sleepy Punishers

Are Harsh Punishers: Daylight Saving Time and Legal Sentences," *Psychological Science* 28, no. 2 (2017): 242–47; for another view of this research, see Holger, Spamann, "Are Sleepy Punishers Really Harsh Punishers? Comment on Cho, Barnes, and Guanara (2017)." *Psychological Science* 29, no. 6 (2018): 1006-1009.

16. Shai Danziger, Jonathan Levav, and Liora Avnaim-Pesso, "Extraneous Factors in Judicial Decisions," *Proceedings of the National Academy of Sciences* 108, no. 17 (2011): 6889–92. Some have criticized this study. See, e.g., Keren Weinshall-Margel and John Shapard, "Overlooked Factors in the Analysis of Parole Decisions." *Proceedings of the National Academy of Sciences* 108, no. 42 (2011): E833-E833. But the authors have offered a persuasive response. See, Shai Danziger, Jonathan Levav, and Liora Avnaim-Pesso, "Reply to Weinshall-Margel and Shapard: Extraneous Factors in Judicial Decisions Persist." *Proceedings of the National Academy of Sciences* 108, no. 42 (2011): E834-E834.

17. Atsunori Ariga and Alejandro Lleras, "Brief and Rare Mental 'Breaks' Keep You Focused: Deactivation and Reactivation of Task Goals Preempt Vigilance Decrements," *Cognition* 118, no. 3 (2011): 439–43.

18. Emily M. Hunter and Cindy Wu, "Give Me a Better Break: Choosing Workday Break Activities to Maximize Resource Recovery," *Journal of Applied Psychology* 101, no. 2 (2016): 302–11.

19. Hannes Zacher, Holly A. Brailsford, and Stacey L. Parker, "Micro-Breaks Matter: A Diary Study on the Effects of Energy Management Strategies on Occupational Well-Being," *Journal of Vocational Behavior* 85, no. 3 (2014): 287–97.

20. Audrey Bergouignan et al., "Effect of Frequent Interruptions of Prolonged Sitting on Self-Perceived Levels of Energy, Mood, Food Cravings and Cognitive Function," *International Journal of Behavioral Nutrition and Physical Activity* 13, no. 1 (2016): 13–24.

21. Li-Ling Wu et al., "Effects of an 8-Week Outdoor Brisk Walking Program on Fatigue in Hi-Tech Industry Employees: A Randomized Control Trial," *Workplace Health & Safety* 63, no. 10 (2015): 436–45; Marily Oppezzo and Daniel L. Schwartz, "Give Your Ideas Some Legs: The Positive Effect of Walking on Creative Thinking," *Journal of Experimental Psychology: Learning, Memory, and Cognition* 40, no. 4 (2014): 1142–52.

22. Johannes Wendsche et al., "Rest Break Organization in Geriatric Care and Turnover: A Multimethod Cross-Sectional Study," *International Journal of Nursing Studies* 51, no. 9 (2014): 1246–57.

23. Sooyeol Kim, Young Ah Park, and Qikun Niu, "Micro-Break Activities at Work to Recover from Daily Work Demands," *Journal of Organizational Behavior* 38, no. 1 (2016): 28–41.

24. Kristen M. Finkbeiner, Paul N. Russell, and William S. Helton, "Rest Improves Performance, Nature Improves Happiness: Assessment of Break Periods on the Abbreviated Vigilance Task," *Consciousness and Cognition* 42 (2016): 277–85.

25. Jo Barton and Jules Pretty, "What Is the Best Dose of Nature and Green Exercise for Improving Mental Health? A Multi-Study Analysis," *Environmental Science & Technology* 44, no. 10 (2010): 3947–55.

26. Elizabeth K. Nisbet and John M. Zelenski, "Underestimating Nearby Nature: Affective Forecasting Errors Obscure the Happy Path to Sustainability," *Psychological Science* 22, no. 9 (2011): 1101–6; Kristen M. Finkbeiner, Paul N. Russell, and William S. Helton, "Rest Improves Performance, Nature Improves Happiness: Assessment of Break Periods on the Abbreviated Vigilance Task," *Consciousness and Cognition* 42 (2016), 277–85.

27. Sooyeol Kim, Young Ah Park, and Qikun Niu, "Micro-Break Activities at Work to Recover from Daily Work Demands," *Journal of Organizational Behavior* 38, no. 1 (2016): 28–41.

28. Hongjai Rhee and Sudong Kim, "Effects of Breaks on Regaining Vitality at Work: An Empirical Comparison of 'Conventional' and 'Smartphone' Breaks," *Computers in Human Behavior* 57 (2016): 160–67.

29. Marjaana Sianoja et al., "Recovery During Lunch Breaks: Testing Long-Term Relations with Energy Levels at Work," *Scandinavian Journal of Work and Organizational Psychology* 1, no. 1 (2016): 1–12.

30. See, e.g., Megan A. McCrory, "Meal Skipping and Variables Related to Energy Balance in Adults: A Brief Review, with Emphasis on the Breakfast Meal," *Physiology & Behavior* 134 (2014): 51–54; and Hania Szajewska and Marek Ruszczyński, "Systematic Review Demonstrating That Breakfast Consumption Influences Body Weight Outcomes in Children and Adolescents in Europe," *Critical Reviews in Food Science and Nutrition* 50, no. 2 (2010): 113–19, where the authors caution that the "results should be interpreted with a substantial degree of caution because of poor reporting of the review process and a lack of information on the quality of the included studies."

31. Emily J. Dhurandhar et al., "The Effectiveness of Breakfast Recommendations on Weight Loss: A Randomized Controlled Trial," *American Journal of Clinical Nutrition* 100, no. 2 (2014): 507–13.

32. Andrew W. Brown, Michelle M. Bohan Brown, and David B. Allison, "Belief Beyond the Evidence: Using the Proposed Effect of Breakfast on Obesity to Show 2 Practices That Distort Scientific Evidence," *American Journal of Clinical Nutrition* 98, no. 5 (2013): 1298–1308; David A. Levitsky and Carly R. Pacanowski, "Effect of Skipping Breakfast on Subsequent Energy Intake," *Physiology & Behavior* 119 (2013): 9–16.

33. Enhad Chowdhury and James Betts, "Should I Eat Breakfast? Health Experts on Whether It Really Is the Most Important Meal of the Day," *Independent*, February 15, 2016. See also Dara Mohammadi, "Is Breakfast Really the Most Important Meal of the Day?" *New Scientist,* March 22, 2016.

34. See, e.g., http://saddesklunch.com, the source of the paragraph's perhaps dubious 62 percent figure.

35. Marjaana Sianoja et al., "Recovery During Lunch Breaks: Testing Long-Term Relations with Energy Levels at Work," *Scandinavian Journal of Work and Organizational Psychology* 1, no. 1 (2016): 1–12.

36. Kevin M. Kniffin et al., "Eating Together at the Firehouse: How Workplace Commensality Relates to the Performance of Firefighters," *Human Performance* 28, no. 4 (2015): 281–306.

37. John P. Trougakos et al., "Lunch Breaks Unpacked: The Role of Autonomy as a Moderator of Recovery During Lunch," *Academy of Management Journal* 57, no. 2 (2014): 405–21.

38. Marjaana Sianoja et al., "Recovery During Lunch Breaks: Testing Long-Term Relations with Energy Levels at Work," *Scandinavian Journal of Work and Organizational Psychology* 1, no. 1 (2016): 1–12. See also Hongjai Rhee and Sudong Kim, "Effects of Breaks on Regaining Vitality at Work: An Empirical Comparison of 'Conventional' and 'Smartphone' Breaks," *Computers in Human Behavior* 57 (2016): 160–67.

39. Wallace Immen, "In This Office, Desks Are for Working, Not Eating Lunch," *Globe and Mail,* February 27, 2017.

40. Mark R. Rosekind et al., "Crew Factors in Flight Operations 9: Effects of Planned Cockpit Rest on Crew Performance and Alertness in Long-Haul Operations," *NASA Technical Reports Server*, 1994, available at https://ntrs.nasa.gov/search.jsp?R=19950006379.

41. Tracey Leigh Signal et al., "Scheduled Napping as a Countermeasure to Sleepiness in Air Traffic Controllers," *Journal of Sleep Research* 18, no. 1 (2009): 11–19.

42. Sergio Garbarino et al., "Professional Shift-Work Drivers Who Adopt Prophylactic Naps Can Reduce the Risk of Car Accidents During Night Work," *Sleep* 27, no. 7 (2004): 1295–1302.

43. Felipe Beijamini et al., "After Being Challenged by a Video Game Problem, Sleep Increases the Chance to Solve It," *PloS ONE* 9, no. 1 (2014): e84342.

44. Bryce A. Mander et al., "Wake Deterioration and Sleep Restoration of Human Learning," *Current Biology* 21, no. 5 (2011): R183–84; Felipe Beijamini et al., "After Being Challenged by a Video Game Problem, Sleep Increases the Chance to Solve It," *PloS ONE* 9, no. 1 (2014): e84342.

45. Nicole Lovato and Leon Lack, "The Effects of Napping on Cognitive Functioning," *Progress in Brain Research* 185 (2010): 155–66; Sara Studte, Emma Bridger, and Axel Mecklinger, "Nap Sleep Preserves Associative but Not Item Memory Performance," *Neurobiology of Learning and Memory* 120 (2015): 84–93.

46. Catherine E. Milner and Kimberly A. Cote, "Benefits of Napping in Healthy Adults: Impact of Nap Length, Time of Day, Age, and Experience with Napping," *Journal of Sleep Research* 18, no. 2 (2009): 272–81; Scott S. Campbell et al., "Effects of a Month-Long Napping Regimen in Older Individuals," *Journal of the American Geriatrics Society* 59, no. 2 (2011): 224–32; Junxin Li et al., "Afternoon Napping and Cognition in Chinese Older Adults: Findings from the China Health and Retirement Longitudinal Study Baseline Assessment," *Journal of the American Geriatrics Society* 65, no. 2 (2016): 373–80.

47. Catherine E. Milner and Kimberly A. Cote, "Benefits of Napping in Healthy Adults: Impact of Nap Length, Time of Day, Age, and Experience with Napping," *Journal of Sleep Research* 18, no. 2 (2009): 272–81.

48. This is especially true when coupled with bright light. See Kosuke Kaida, Yuji Takeda, and Kazuyo Tsuzuki, "The Relationship Between Flow, Sleepiness and Cognitive Performance: The Effects of Short Afternoon Nap and Bright Light Exposure," *Industrial Health* 50, no. 3 (2012): 189–96.

49. Nicholas Bakalar, "Regular Midday Snoozes Tied to a Healthier Heart," *New York Times*, February 13, 2007, reporting on Androniki Naska et al., "Siesta in Healthy Adults and Coronary Mortality in the General Population," *Archives of Internal Medicine* 167, no. 3 (2007): 296–301. Cautionary note: This study showed a correlation between napping and the reduced risk of heart disease, not necessarily that napping caused the health benefit.

50. Brice Faraut et al., "Napping Reverses the Salivary Interleukin-6 and Urinary Norepinephrine Changes Induced by Sleep Restriction," *Journal of Clinical Endocrinology & Metabolism* 100, no. 3 (2015): E416–26.

51. Mohammad Zaregarizi et al., "Acute Changes in Cardiovascular Function During the Onset Period of Daytime Sleep: Comparison to Lying Awake and Standing," *Journal of Applied Physiology* 103, no. 4 (2007): 1332–38.

52. Amber Brooks and Leon C. Lack, "A Brief Afternoon Nap Following Nocturnal Sleep

Restriction: Which Nap Duration Is Most Recuperative?" *Sleep* 29, no. 6 (2006): 831–40.

53. Amber J. Tietzel and Leon C. Lack, "The Recuperative Value of Brief and Ultra-Brief Naps on Alertness and Cognitive Performance," *Journal of Sleep Research* 11, no. 3 (2002): 213–18.

54. Catherine E. Milner and Kimberly A. Cote, "Benefits of Napping in Healthy Adults: Impact of Nap Length, Time of Day, Age, and Experience with Napping," *Journal of Sleep Research* 18, no. 2 (2009): 272–81.

55. Luise A. Reyner and James A. Horne, "Suppression of Sleepiness in Drivers: Combination of Caffeine with a Short Nap," *Psychophysiology* 34, no. 6 (1997): 721–25.

56. Mitsuo Hayashi, Akiko Masuda, and Tadao Hori, "The Alerting Effects of Caffeine, Bright Light and Face Washing After a Short Daytime Nap," *Clinical Neurophysiology* 114, no. 12 (2003): 2268–78.

57. Renwick McLean, "For Many in Spain, Siesta Ends," *New York Times,* January 1, 2006; Jim Yardley, "Spain, Land of 10 P.M. Dinners, Asks If It's Time to Reset Clock," *New York Times,* February 17, 2014; Margarita Mayo, "Don't Call It the 'End of the Siesta': What Spain's New Work Hours Really Mean, *Harvard Business Review,* April 13, 2016.

58. Ahmed S. BaHammam, "Sleep from an Islamic Perspective," *Annals of Thoracic Medicine* 6, no. 4 (2011): 187–92.

59. Dan Bilefsky and Christina Anderson, "A Paid Hour a Week for Sex? Swedish Town Considers It," *New York Times,* February 23, 2017.

CHAPTER 2. TIME HACKER'S HANDBOOK

1. Mayo Clinic staff, "Napping: Do's and Don'ts for Healthy Adults," available at https://www.mayoclinic.org/healthy-lifestyle/adult-health/in-depth/napping/art-20048319.

2. Hannes Zacher, Holly A. Brailsford, and Stacey L. Parker, "Micro-Breaks Matter: A Diary Study on the Effects of Energy Management Strategies on Occupational Well-Being," *Journal of Vocational Behavior* 85, no. 3 (2014): 287–97.

3. Daniel Z. Levin, Jorge Walter, and J. Keith Murnighan, "The Power of Reconnection: How Dormant Ties Can Surprise You," *MIT Sloan Management Review* 52, no. 3 (2011): 45–50.

4. Christopher Peterson et al., "Strengths of Character, Orientations to Happiness, and Life Satisfaction," *Journal of Positive Psychology* 2, no. 3 (2007): 149–56.

5. See Anna Brones and Johanna Kindvall, *Fika: The Art of the Swedish Coffee Break* (Berkeley, CA: Ten Speed Press, 2015); Anne Quito, "This Four-Letter Word Is the Swedish Key to Happiness at Work," *Quartz,* March 14, 2016.

6. Charlotte Fritz, Chak Fu Lam, and Gretchen M. Spreitzer, "It's the Little Things That Matter: An Examination of Knowledge Workers' Energy Management," *Academy of Management Perspectives* 25, no. 3 (2011): 28–39.

7. Lesley Alderman, "Breathe. Exhale. Repeat: The Benefits of Controlled Breathing," *New York Times,* November 9, 2016.

8. Kristen M. Finkbeiner, Paul N. Russell, and William S. Helton, "Rest Improves Performance, Nature Improves Happiness: Assessment of Break Periods on the Abbreviated Vigilance Task," *Consciousness and Cognition* 42 (2016): 277–85.

9. Angela Duckworth, *Grit: The Power of Passion and Perseverance* (New York: Scribner, 2016), 118.

10. Stephanie Pappas, "As Schools Cut Recess, Kids' Learning Will Suffer, Experts Say," *Live Science* (2011). Available at https://www.livescience.com/15555-schools-cut-recess-learning-suffers.html.

11. Claude Brodesser-Akner, "Christie: 'Stupid' Mandatory Recess Bill Deserved My Veto," NJ.com, January 20, 2016, available at http://www.nj.com/politics/index.ssf/2016/01/christie_stupid_law_assuring_kids_recess_deserved.html.

12. Olga S. Jarrett et al., "Impact of Recess on Classroom Behavior: Group Effects and Individual Differences," *Journal of Educational Research* 92, no. 2 (1998): 121–26.

13. Catherine N. Rasberry et al., "The Association Between School-Based Physical Activity, Including Physical Education, and Academic Performance: A Systematic Review of the Literature," *Preventive Medicine* 52 (2011): S10–20.

14. Romina M. Barros, Ellen J. Silver, and Ruth E. K. Stein, "School Recess and Group Classroom Behavior," *Pediatrics* 123, no. 2 (2009): 431–36; Anthony D. Pellegrini and Catherine M. Bohn, "The Role of Recess in Children's Cognitive Performance and School Adjustment," *Educational Researcher* 34, no. 1 (2005): 13–19.

15. Sophia Alvarez Boyd, "Not All Fun and Games: New Guidelines Urge Schools to Rethink Recess," National Public Radio, February 1, 2017.

16. Timothy D. Walker, "How Kids Learn Better by Taking Frequent Breaks Throughout the Day," *KQED News Mind Shift*, April 18, 2017; Christopher Connelly, "More Playtime! How Kids Succeed with Recess Four Times a Day at School," *KQED News*, January 4, 2016.

CHAPTER 3. BEGINNINGS

1. Anne G. Wheaton, Gabrielle A. Ferro, and Janet B. Croft, "School Start Times for Middle School and High School Students—United States, 2011–12 School Year," *Morbidity and Mortality Weekly Report* 64, no. 3 (August 7, 2015): 809–13.

2. Karen Weintraub, "Young and Sleep Deprived," *Monitor on Psychology* 47, no. 2 (2016): 46, citing Katherine M. Keyes et al., "The Great Sleep Recession: Changes in Sleep Duration Among US Adolescents, 1991–2012," *Pediatrics* 135, no. 3 (2015): 460–68.

3. Finley Edwards, "Early to Rise? The Effect of Daily Start Times on Academic Performance," *Economics of Education Review* 31, no. 6 (2012): 970–83.

4. Reut Gruber et al., "Sleep Efficiency (But Not Sleep Duration) of Healthy School-Age Children Is Associated with Grades in Math and Languages," *Sleep Medicine* 15, no. 12 (2014): 1517–25.

5. Adolescent Sleep Working Group, "School Start Times for Adolescents," *Pediatrics* 134, no. 3 (2014): 642–49.

6. Kyla Wahlstrom et al., "Examining the Impact of Later High School Start Times on the Health and Academic Performance of High School Students: A Multi-Site Study," Center for Applied Research and Educational Improvement (2014). See also Robert Daniel Vorona et al., "Dissimilar Teen Crash Rates in Two Neighboring Southeastern Virginia Cities with Different High School Start Times," *Journal of Clinical Sleep Medicine* 7, no. 2 (2011): 145–51.

7. Pamela Malaspina McKeever and Linda Clark, "Delayed High School Start Times Later than 8:30 AM and Impact on Graduation Rates and Attendance Rates," *Sleep Health* 3, no. 2 (2017): 119–25; Carolyn Crist, "Later School Start Times Catch on Nationwide," *District Administrator*, March 28, 2017.

8. Anne G. Wheaton, Daniel P. Chapman, and Janet B. Croft, "School Start Times, Sleep, Behavioral, Health, and Academic Outcomes: A Review of the Literature," *Journal of School Health* 86, no. 5 (2016): 363–81.

9. Judith A. Owens, Katherine Belon, and Patricia Moss, "Impact of Delaying School Start Time on Adolescent Sleep, Mood, and Behavior," *Archives of Pediatrics & Adolescent Medicine* 164, no. 7 (2010): 608–14; Nadine Perkinson-Gloor, Sakari Lemola, and Alexander Grob, "Sleep Duration, Positive Attitude Toward Life, and Academic Achievement: The Role of Daytime Tiredness, Behavioral Persistence, and School Start Times," *Journal of Adolescence* 36, no. 2 (2013): 311–18; Timothy I. Morgenthaler et al., "High School Start Times and the Impact on High School Students: What We Know, and What We Hope to Learn," *Journal of Clinical Sleep Medicine* 12, no. 12 (2016): 168–89; Julie Boergers, Christopher J. Gable, and Judith A. Owens, "Later School Start Time Is Associated with Improved Sleep and Daytime Functioning in Adolescents," *Journal of Developmental & Behavioral Pediatrics* 35, no. 1 (2014): 11–17; Kyla Wahlstrom, "Changing Times: Findings from the First Longitudinal Study of Later High School Start Times," *NASSP Bulletin* 86, no. 633 (2002): 3–21; Dubi Lufi, Orna Tzischinsky, and Stav Hadar, "Delaying School Starting Time by One Hour: Some Effects on Attention Levels in Adolescents," *Journal of Clinical Sleep Medicine* 7, no. 2 (2011): 137–43.

10. Scott E. Carrell, Teny Maghakian, and James E. West, "A's from Zzzz's? The Causal Effect of School Start Time on the Academic Achievement of Adolescents," *American Economic Journal: Economic Policy* 3, no. 3 (2011): 62–81.

11. M.D.R. Evans, Paul Kelley, and Johnathan Kelley, "Identifying the Best Times for Cognitive Functioning Using New Methods: Matching University Times to Undergraduate Chronotypes," *Frontiers in Human Neuroscience* 11 (2017): 188.

12. Finley Edwards, "Early to Rise? The Effect of Daily Start Times on Academic Performance," *Economics of Education Review* 31, no. 6 (2012): 970–83.

13. Brian A. Jacob and Jonah E. Rockoff, "Organizing Schools to Improve Student Achievement: Start Times, Grade Configurations, and Teacher Assignments," *Education Digest* 77, no. 8 (2012): 28–34.

14. Anne G. Wheaton, Gabrielle A. Ferro, and Janet B. Croft, "School Start Times for Middle School and High School Students—United States, 2011–12 School Year," *Morbidity and Mortality Weekly Report* 64, no. 30 (August 7, 2015): 809–13; Karen Weintraub, "Young and Sleep Deprived," *Monitor on Psychology* 47, no. 2 (2016): 46.

15. The term originally came from Michael S. Shum, "The Role of Temporal Landmarks in Autobiographical Memory Processes," *Psychological Bulletin* 124, no. 3 (1998): 423. Shum, who received a PhD in psychology from Northwestern University, left behavioral science, earned a second PhD in English, and is now a novelist.

16. Hengchen Dai, Katherine L. Milkman, and Jason Riis, "The Fresh Start Effect: Temporal Landmarks Motivate Aspirational Behavior," *Management Science* 60, no. 10 (2014): 2563–82. In 2018, California Governor Jerry Brown vetoed legislation that would have delayed start times for most of the state's middle and high schools until

8:30 a.m. Mini Racker, "California Gov. Jerry Brown Rejects Bill to Prohibit Schools from Starting Before 8:30 a.m." *Los Angeles Times*, Sep. 20, 2018.

17. Ibid.

18. Johanna Peetz and Anne E. Wilson, "Marking Time: Selective Use of Temporal Landmarks as Barriers Between Current and Future Selves," *Personality and Social Psychology Bulletin* 40, no. 1 (2014): 44–56.

19. Hengchen Dai, Katherine L. Milkman, and Jason Riis, "The Fresh Start Effect: Temporal Landmarks Motivate Aspirational Behavior," *Management Science* 60, no. 10 (2014): 2563–82.

20. Jason Riis, "Opportunities and Barriers for Smaller Portions in Food Service: Lessons from Marketing and Behavioral Economics," *International Journal of Obesity* 38 (2014): S19–24.

21. Hengchen Dai, Katherine L. Milkman, and Jason Riis, "The Fresh Start Effect: Temporal Landmarks Motivate Aspirational Behavior," *Management Science* 60, no. 10 (2014): 2563–82.

22. Sadie Stein, "I Always Start on 8 January," *Paris Review*, January 8, 2013; Alison Beard, "Life's Work: An Interview with Isabel Allende," *Harvard Business Review*, May 2016.

23. Hengchen Dai, Katherine L. Milkman, and Jason Riis, "Put Your Imperfections Behind You: Temporal Landmarks Spur Goal Initiation When They Signal New Beginnings," *Psychological Science* 26, no. 12 (2015): 1927–36.

24. Jordi Brandts, Christina Rott, and Carles Solà, "Not Just Like Starting Over: Leadership and Revivification of Cooperation in Groups," *Experimental Economics* 19, no. 4 (2016): 792–818.

25. Jason Riis, "Opportunities and Barriers for Smaller Portions in Food Service: Lessons from Marketing and Behavioral Economics," *International Journal of Obesity* 38 (2014): S19–24.

26. John C. Norcross, Marci S. Mrykalo, and Matthew D. Blagys, "Auld Lang Syne: Success Predictors, Change Processes, and Self-Reported Outcomes of New Year's Resolvers and Nonresolvers," *Journal of Clinical Psychology* 58, no. 4 (2002): 397–405.

27. Lisa B. Kahn, "The Long-Term Labor Market Consequences of Graduating from College in a Bad Economy," *Labour Economics* 17, no. 2 (2010): 303–16.

28. This idea is a cornerstone of chaos and complexity theory. See, e.g., Dean Rickles, Penelope Hawe, and Alan Shiell, "A Simple Guide to Chaos and Complexity," *Journal of Epidemiology & Community Health* 61, no. 11 (2007): 933–37.

29. Philip Oreopoulos, Till von Wachter, and Andrew Heisz, "The Short- and Long-Term Career Effects of Graduating in a Recession," *American Economic Journal: Applied Economics* 4, no. 1 (2012): 1–29.

30. Antoinette Schoar and Luo Zuo, "Shaped by Booms and Busts: How the Economy Impacts CEO Careers and Management Styles," *Review of Financial Studies* (forthcoming). Available at SSRN: https://ssrn.com/abstract=1955612 or http://dx.doi.org/10.2139/ssrn.1955612.

31. Paul Oyer, "The Making of an Investment Banker: Stock Market Shocks, Career Choice, and Lifetime Income," *Journal of Finance* 63, no. 6 (2008): 2601–28.

32. Joseph G. Altonji, Lisa B. Kahn, and Jamin D. Speer, "Cashier or Consultant? Entry Labor Market Conditions, Field of Study, and Career Success," *Journal of Labor Economics* 34, no. S1 (2016): S361–401.

33. Jaison R. Abel, Richard Deitz, and Yaqin Su, "Are Recent College Graduates Finding Good Jobs?" *Current Issues in Economics and Finance* 20, no. 1 (2014).

34. Paul Beaudry and John DiNardo, "The Effect of Implicit Contracts on the Movement of Wages over the Business Cycle: Evidence from Micro Data," *Journal of Political Economy* 99, no. 4 (1991): 665–88; see also Darren Grant, "The Effect of Implicit Contracts on the Movement of Wages over the Business Cycle: Evidence from the National Longitudinal Surveys," *ILR Review* 56, no. 3 (2003): 393–408.

35. David P. Phillips and Gwendolyn E. C. Barker, "A July Spike in Fatal Medication Errors: A Possible Effect of New Medical Residents," *Journal of General Internal Medicine* 25, no. 8 (2010): 774–79.

36. Michael J. Englesbe et al., "Seasonal Variation in Surgical Outcomes as Measured by the American College of Surgeons-National Surgical Quality Improvement Program (ACS-NSQIP)," *Annals of Surgery* 246, no. 3 (2007): 456–65

37. David L. Olds et al., "Effect of Home Visiting by Nurses on Maternal and Child Mortality: Results of a 2-Decade Follow-up of a Randomized Clinical Trial," *JAMA Pediatrics* 168, no. 9 (2014): 800–806; David L. Olds et al., "Effects of Home Visits by Paraprofessionals and by Nurses on Children: Follow-up of a Randomized Trial at Ages 6 and 9 Years," *JAMA Pediatrics* 168, no. 2 (2014): 114–21; Sabrina Tavernise, "Visiting Nurses, Helping Mothers on the Margins," *New York Times,* March 8, 2015.

38. David L. Olds, Lois Sadler, and Harriet Kitzman, "Programs for Parents of Infants and Toddlers: Recent Evidence from Randomized Trials," *Journal of Child Psychology and Psychiatry* 48, no. 3–4 (2007): 355–91; William Thorland et al., "Status of Breastfeeding and Child Immunization Outcomes in Clients of the Nurse-Family Partnership," *Maternal and Child Health Journal* 21, no. 3 (2017): 439–45; Nurse-Family Partnership, "Trials and Outcomes" (2017). Available at http://www.nursefamily partnership.org/proven-results/published-research.

CHAPTER 3. TIME HACKER'S HANDBOOK

1. Gary Klein, "Performing a Project Premortem," *Harvard Business Review* 85, no. 9 (2007): 18–19.

2. Marc Meredith and Yuval Salant, "On the Causes and Consequences of Ballot Order Effects," *Political Behavior* 35, no. 1 (2013): 175–97; Darren P. Grant, "The Ballot Order Effect Is Huge: Evidence from Texas," May 9, 2016. Available at https://ssrn .com/abstract=2777761.

3. Shai Danziger, Jonathan Levav, and Liora Avnaim-Pesso, "Extraneous Factors in Judicial Decisions," *Proceedings of the National Academy of Sciences* 108, no. 17 (2011): 6889–92.

4. Antonia Mantonakis et al., "Order in Choice: Effects of Serial Position on Preferences," *Psychological Science* 20, no. 11 (2009): 1309–12.

5. Uri Simonsohn and Francesca Gino, "Daily Horizons: Evidence of Narrow Bracketing in Judgment from 10 Years of MBA Admissions Interviews," *Psychological Science* 24, no. 2 (2013): 219–24.

6. Shai Danziger, Jonathan Levav, and Liora Avnaim-Pesso. "Extraneous Factors in Judicial Decisions," *Proceedings of the National Academy of Sciences* 108, no. 17 (2011): 6889–92.

7. Lionel Page and Katie Page, "Last Shall Be First: a Field Study of Biases in Sequential

Performance Evaluation on the Idol Series," *Journal of Economic Behavior & Organization* 73, no. 2 (2010): 186–98; Adam Galinsky and Maurice Schweitzer, *Friend & Foe: When to Cooperate, When to Compete, and How to Succeed at Both* (New York: Crown Business, 2015), 229.

8. Wändi Bruine de Bruin, "Save the Last Dance for Me: Unwanted Serial Position Effects in Jury Evaluations," *Acta Psychologica* 118, no. 3 (2005): 245–60.

9. Steve Inskeep and Shankar Vedantan, "Deciphering Hidden Biases During Interviews," National Public Radio's *Morning Edition*, March 6, 2013, interview with Uri Simonsohn, citing Uri Simonsohn and Francesca Gino, "Daily Horizons: Evidence of Narrow Bracketing in Judgment from 10 Years of MBA Admissions Interviews," *Psychological Science* 24, no. 2 (2013): 219–24.

10. Michael Watkins, *The First 90 Days: Critical Success Strategies for New Leaders at All Levels*, read by Kevin T. Norris (Flushing, NY: Gildan Media LLC, 2013). Audiobook.

11. Ram Charan, Stephen Drotter, and James Noel, *The Leadership Pipeline: How to Build the Leadership Powered Company*, 2nd ed. (San Francisco: Jossey-Bass, 2011).

12. Harrison Wellford, "Preparing to Be President on Day One," *Public Administration Review* 68, no. 4 (2008): 618–23.

13. Corinne Bendersky and Neha Parikh Shah, "The Downfall of Extraverts and the Rise of Neurotics: The Dynamic Process of Status Allocation in Task Groups," *Academy of Management Journal* 56, no. 2 (2013): 387–406.

14. Brian J. Fogg, "A Behavior Model for Persuasive Design" in *Proceedings of the 4th International Conference on Persuasive Technology* (New York: ACM, 2009). For an explanation of motivational waves, see https://www.youtube.com/watch?v=fqUSjHjIEFg.

15. Karl E. Weick, "Small Wins: Redefining the Scale of Social Problems," *American Psychologist* 39, no. 1 (1984): 40–49.

16. Teresa Amabile and Steven Kramer, *The Progress Principle: Using Small Wins to Ignite Joy, Engagement, and Creativity at Work* (Cambridge, MA: Harvard Business Review Press, 2011).

17. Nicholas Wolfinger, "Want to Avoid Divorce? Wait to Get Married, but Not Too Long," *Institute for Family Studies*, July 16, 2015, which analyzed data from Casey E. Copen et al., "First Marriages in the United States: Data from the 2006–2010 National Survey of Family Growth," *National Health Statistics Reports*, no. 49, March 22, 2012.

18. Scott Stanley et al., "Premarital Education, Marital Quality, and Marital Stability: Findings from a Large, Random Household Survey," *Journal of Family Psychology* 20, no. 1 (2006): 117–26.

19. Andrew Francis-Tan and Hugo M. Mialon, "'A Diamond Is Forever' and Other Fairy Tales: The Relationship Between Wedding Expenses and Marriage Duration," *Economic Inquiry* 53, no. 4 (2015): 1919–30.

CHAPTER 4. MIDPOINTS

1. Elliot Jaques, "Death and the Mid-Life Crisis," *International Journal of Psycho-Analysis* 46 (1965): 502–14.

2. The popularity was helped along by Gail Sheehy, author of the blockbuster 1974 book *Passages: Predictable Crises of Adult Life*, which describes variations of the midlife crisis but does not credit Jaques until page 369.

3. Elliot Jaques, "Death and the Mid-Life Crisis," *International Journal of Psycho-Analysis* 46 (1965): 502–14.

4. Arthur A. Stone et al., "A Snapshot of the Age Distribution of Psychological Well-Being in the United States," *Proceedings of the National Academy of Sciences* 107, no. 22 (2010): 9985–90.

5. David G. Blanchflower and Andrew J. Oswald, "Is Well-Being U-Shaped over the Life Cycle?" *Social Science & Medicine* 66, no. 8 (2008): 1733–49.

6. See also Terence Chai Cheng, Nattavudh Powdthavee, and Andrew J. Oswald, "Longitudinal Evidence for a Midlife Nadir in Human Well-Being: Results from Four Data Sets," *Economic Journal* 127, no. 599 (2017): 126–42; Andrew Steptoe, Angus Deaton, and Arthur A. Stone, "Subjective Wellbeing, Health, and Ageing," *Lancet* 385, no. 9968 (2015): 640–48; Paul Frijters and Tony Beatton, "The Mystery of the U-Shaped Relationship Between Happiness and Age," *Journal of Economic Behavior & Organization* 82, no. 2–3 (2012): 525–42; Carol Graham, *Happiness Around the World: The Paradox of Happy Peasants and Miserable Millionaires* (Oxford: Oxford University Press, 2009). Some research has shown that while the U-shape remains consistent across countries, it varies from nation to nation in the "turning point"—when well-being reaches its nadir and begins its ascent. See Carol Graham and Julia Ruiz Pozuelo, "Happiness, Stress, and Age: How the U-Curve Varies Across People and Places," *Journal of Population Economics* 30, no. 1 (2017): 225–64; Bert van Landeghem, "A Test for the Convexity of Human Well-Being over the Life Cycle: Longitudinal Evidence from a 20-Year Panel," *Journal of Economic Behavior & Organization* 81, no. 2 (2012): 571–82.

7. David G. Blanchflower and Andrew J. Oswald, "Is Well-Being U-Shaped over the Life Cycle?" *Social Science & Medicine* 66, no. 8 (2008): 1733–49.

8. Hannes Schwandt, "Unmet Aspirations as an Explanation for the Age U-Shape in Wellbeing," *Journal of Economic Behavior & Organization* 122 (2016): 75–87.

9. Alexander Weiss et al., "Evidence for a Midlife Crisis in Great Apes Consistent with the U-Shape in Human Well-Being," *Proceedings of the National Academy of Sciences* 109, no. 49 (2012): 19949–52.

10. Maferima Touré-Tillery and Ayelet Fishbach, "The End Justifies the Means, but Only in the Middle," *Journal of Experimental Psychology: General* 141, no. 3 (2012): 570–83.

11. Ibid.

12. Niles Eldredge and Stephen Jay Gould, "Punctuated Equilibria: An Alternative to Phyletic Gradualism," in Thomas Schopf, ed., *Models in Paleobiology* (San Francisco: Freeman, Cooper and Company, 1972), 82–115; Stephen Jay Gould and Niles Eldredge, "Punctuated Equilibria: The Tempo and Mode of Evolution Reconsidered," *Paleobiology* 3, no. 2 (1977): 115–51.

13. Connie J. G. Gersick, "Time and Transition in Work Teams: Toward a New Model of Group Development," *Academy of Management Journal* 31, no. 1 (1988): 9–41.

14. Ibid.

15. Connie J. G. Gersick, "Marking Time: Predictable Transitions in Task Groups," *Academy of Management Journal* 32, no. 2 (1989): 274–309.

16. Connie J. G. Gersick, "Pacing Strategic Change: The Case of a New Venture," *Academy of Management Journal* 37, no. 1 (1994): 9–45.

17. Malcolm Moran, "Key Role for Coaches in Final," *New York Times*, March 29, 1982;

Jack Wilkinson, "UNC's Crown a Worthy One," *New York Daily News,* March 20, 1982; Curry Kirkpatrick, "Nothing Could Be Finer," *Sports Illustrated,* April 5, 1982.

18. Curry Kirkpatrick, "Nothing Could Be Finer," *Sports Illustrated,* April 5, 1982.

19. Malcolm Moran, "North Carolina Slips Past Georgetown by 63–62," *New York Times,* March 30, 1982.

20. Jonah Berger and Devin Pope, "Can Losing Lead to Winning?" *Management Science* 57, no. 5 (2011): 817–27.

21. Ibid.

22. "Key Moments in Dean Smith's Career," *Charlotte Observer,* February 8, 2015.

CHAPTER 4. TIME HACKER'S HANDBOOK

1. Andrea C. Bonezzi, Miguel Brendl, and Matteo De Angelis, "Stuck in the Middle: The Psychophysics of Goal Pursuit," *Psychological Science* 22, no. 5 (2011): 607–12.

2. See Colleen M. Seifert and Andrea L. Patalano, "Memory for Incomplete Tasks: A Re-Examination of the Zeigarnik Effect," *Proceedings of the Thirteenth Annual Conference of the Cognitive Science Society* (Mahwah, NJ: Lawrence Erlbaum Associates, 1991), 114.

3. Brad Isaac, "Jerry Seinfeld's Productivity Secret," *Lifehacker,* July 24, 2007, 276–86.

4. Adam Grant, *2 Fail-Proof Techniques to Increase Your Productivity* (Inc. Video). Available at https://www.inc.com/adam-grant/productivity-playbook-failproof-productivity -techniques.html.

5. Minjung Koo and Ayelet Fishbach, "Dynamics of Self-Regulation: How (Un) Accom-plished Goal Actions Affect Motivation," *Journal of Personality and Social Psychology* 94, no. 2 (2008): 183–95.

6. Cameron Ford and Diane M. Sullivan, "A Time for Everything: How the Timing of Novel Contributions Influences Project Team Outcomes," *Journal of Organizational Behavior* 25, no. 2 (2004): 279–92.

7. J. Richard Hackman and Ruth Wageman, "A Theory of Team Coaching," *Academy of Management Review* 30, no. 2 (2005): 269–87.

8. Hannes Schwandt, "Why So Many of Us Experience a Midlife Crisis," *Harvard Business Review,* April 20, 2015. Available at https://hbr.org/2015/04/why-so-many-of-us -experience-a-midlife-crisis.

9. Minkyung Koo et al., "It's a Wonderful Life: Mentally Subtracting Positive Events Improves People's Affective States, Contrary to Their Affective Forecasts," *Journal of Personality and Social Psychology* 95, no. 5 (2008): 1217–24.

10. Juliana G. Breines and Serena Chen, "Self-Compassion Increases Self-Improvement Motivation," *Personality and Social Psychology Bulletin* 38, no. 9 (2012): 1133–43; Kris-tin D. Neff and Christopher K. Germer, "A Pilot Study and Randomized Controlled Trial of the Mindful Self-Compassion Program," *Journal of Clinical Psychology* 69, no. 1 (2013): 28–44; Kristin D. Neff, "The Development and Validation of a Scale to Measure Self-Compassion," *Self and Identity* 2, no. 3 (2003): 223–50; Leah B. Shapira and Myriam Mongrain, "The Benefits of Self-Compassion and Optimism Exercises for Individuals Vulnerable to Depression," *Journal of Positive Psychology* 5, no. 5 (2010): 377–89; Lisa M. Yarnell et al., "Meta-Analysis of Gender Differences in Self-Compassion," *Self and Identity* 14, no. 5 (2015): 499–520.

CHAPTER 5. ENDINGS

1. Running USA, *2015 Running USA Annual Marathon Report*, May 25, 2016, available at http://www.runningusa.org/marathon-report-2016; Ahotu Marathons, *2017–2018 Marathon Schedule*, available at http://marathons.ahotu.com/calendar/marathon; Skechers Performance Los Angeles Marathon, *Race History*, available at http://www.lamarathon.com/press/race-history; Andrew Cave and Alex Miller, "Marathon Runners Sign Up in Record Numbers," *Telegraph,* March 24, 2016.

2. Adam L. Alter and Hal E. Hershfield, "People Search for Meaning When They Approach a New Decade in Chronological Age," *Proceedings of the National Academy of Sciences* 111, no. 48 (2014): 17066–70. For a critique of some of Alter and Hershfield's data and conclusions, see Erik G. Larsen, "Commentary On: People Search for Meaning When They Approach a New Decade in Chronological Age," *Frontiers in Psychology* 6 (2015): 792.

3. Adam L. Alter and Hal E. Hershfield, "People Search for Meaning When They Approach a New Decade in Chronological Age," *Proceedings of the National Academy of Sciences* 111, no. 48 (2014): 17066–70.

4. Jim Chairusmi, "When Super Bowl Scoring Peaks—or Timing Your Bathroom Break," *Wall Street Journal*, February 4, 2017.

5. Clark L. Hull, "The Goal-Gradient Hypothesis and Maze Learning," *Psychological Review* 39, no. 1 (1932): 25.

6. Clark L. Hull, "The Rat's Speed-of-Locomotion Gradient in the Approach to Food," *Journal of Comparative Psychology* 17, no. 3 (1934): 393.

7. Arthur B. Markman and C. Miguel Brendl, "The Influence of Goals on Value and Choice," *Psychology of Learning and Motivation* 39 (2000): 97–128; Minjung Koo and Ayelet Fishbach, "Dynamics of Self-Regulation: How (Un)Accomplished Goal Actions Affect Motivation," *Journal of Personality and Social Psychology* 94, no. 2 (2008): 183–95; Andrea Bonezzi, C. Miguel Brendl, and Matteo De Angelis, "Stuck in the Middle: The Psychophysics of Goal Pursuit," *Psychological Science* 22, no. 5 (2011): 607–12; Szu-Chi Huang, Jordan Etkin, and Liyin Jin, "How Winning Changes Motivation in Multiphase Competitions," *Journal of Personality and Social Psychology* 112, no. 6 (2017): 813–37; Kyle E. Conlon et al., "Eyes on the Prize: The Longitudinal Benefits of Goal Focus on Progress Toward a Weight Loss Goal," *Journal of Experimental Social Psychology* 47, no. 4 (2011): 853–55.

8. Kristen Berman, "The Deadline Made Me Do It," *Scientific American,* November 9, 2016, available at https://blogs.scientificamerican.com/mind-guest-blog/the-deadline-made-me-do-it/.

9. John C. Birkimer et al., "Effects of Refutational Messages, Thought Provocation, and Decision Deadlines on Signing to Donate Organs," *Journal of Applied Social Psychology* 24, no. 19 (1994): 1735–61.

10. Suzanne B. Shu and Ayelet Gneezy, "Procrastination of Enjoyable Experiences," *Journal of Marketing Research* 47, no. 5 (2010): 933–44.

11. Uri Gneezy, Ernan Haruvy, and Alvin E. Roth, "Bargaining Under a Deadline: Evidence from the Reverse Ultimatum Game," *Games and Economic Behavior* 45, no. 2 (2003): 347–68; Don A. Moore, "The Unexpected Benefits of Final Deadlines in Negotiation," *Journal of Experimental Social Psychology* 40, no. 1 (2004): 121–27.

12. Szu-Chi Huang and Ying Zhang, "All Roads Lead to Rome: The Impact of Multiple

Attainment Means on Motivation," *Journal of Personality and Social Psychology* 104, no. 2 (2013): 236–48.

13. Teresa M. Amabile, William DeJong, and Mark R. Lepper, "Effects of Externally Imposed Deadlines on Subsequent Intrinsic Motivation," *Journal of Personality and Social Psychology* 34, no. 1 (1976): 92–98; Teresa M. Amabile, "The Social Psychology of Creativity: A Componential Conceptualization," *Journal of Personality and Social Psychology* 45, no. 2 (1983): 357–77; Edward L. Deci and Richard M. Ryan, "The 'What' and 'Why' of Goal Pursuits: Human Needs and the Self-Determination of Behavior," *Psychological Inquiry* 11, no. 4 (2000): 227–68.

14. See, e.g., Marco Pinfari, "Time to Agree: Is Time Pressure Good for Peace Negotiations?" *Journal of Conflict Resolution* 55, no. 5 (2011): 683–709.

15. Ed Diener, Derrick Wirtz, and Shigehiro Oishi, "End Effects of Rated Life Quality: The James Dean Effect," *Psychological Science* 12, no. 2 (2001): 124–28.

16. Daniel Kahneman et al., "When More Pain Is Preferred to Less: Adding a Better End," *Psychological Science* 4, no. 6 (1993): 401–405; Barbara L. Fredrickson and Daniel Kahneman, "Duration Neglect in Retrospective Evaluations of Affective Episodes," *Journal of Personality and Social Psychology* 65, no. 1 (1993): 45–55; Charles A. Schreiber and Daniel Kahneman, "Determinants of the Remembered Utility of Aversive Sounds," *Journal of Experimental Psychology: General* 129, no. 1 (2000): 27–42.

17. Donald A. Redelmeier and Daniel Kahneman, "Patients' Memories of Painful Medical Treatments: Real-Time and Retrospective Evaluations of Two Minimally Invasive Procedures," *Pain* 66, no. 1 (1996): 3–8.

18. Daniel Kahneman, *Thinking, Fast and Slow* (New York: Farrar, Straus and Giroux, 2011), 380.

19. George F. Loewenstein and Dražen Prelec, "Preferences for Sequences of Outcomes," *Psychological Review* 100, no. 1 (1993): 91–108; Hans Baumgartner, Mita Sujan, and Dan Padgett, "Patterns of Affective Reactions to Advertisements: The Integration of Moment-to-Moment Responses into Overall Judgments," *Journal of Marketing Research* 34, no. 2 (1997): 219–32; Amy M. Do, Alexander V. Rupert, and George Wolford, "Evaluations of Pleasurable Experiences: The Peak-End Rule," *Psychonomic Bulletin & Review* 15, no. 1 (2008): 96–98.

20. Andrew Healy and Gabriel S. Lenz, "Substituting the End for the Whole: Why Voters Respond Primarily to the Election-Year Economy," *American Journal of Political Science* 58, no. 1 (2014): 31–47; Andrews Healy and Neil Malhotra, "Myopic Voters and Natural Disaster Policy," *American Political Science Review* 103, no. 3 (2009): 387–406.

21. George E. Newman, Kristi L. Lockhart, and Frank C. Keil, "'End-of-Life' Biases in Moral Evaluations of Others," *Cognition* 115, no. 2 (2010): 343–49.

22. Ibid.

23. Tammy English and Laura L. Carstensen, "Selective Narrowing of Social Networks Across Adulthood Is Associated with Improved Emotional Experience in Daily Life," *International Journal of Behavioral Development* 38, no. 2 (2014): 195–202.

24. Laura L. Carstensen, Derek M. Isaacowitz, and Susan T. Charles, "Taking Time Seriously: A Theory of Socioemotional Selectivity," *American Psychologist* 54, no. 3 (1999): 165–81.

25. Ibid.

26. Other research has produced similar findings. See, e.g., Frieder R. Lang, "Endings and Continuity of Social Relationships: Maximizing Intrinsic Benefits Within Personal Networks When Feeling Near to Death," *Journal of Social and Personal Relationships* 17, no. 2 (2000): 155–82; Cornelia Wrzus et al., "Social Network Changes and Life Events Across the Life Span: A Meta-Analysis," *Psychological Bulletin* 139, no. 1 (2013): 53–80.

27. Laura L. Carstensen, Derek M. Isaacowitz, and Susan T. Charles, "Taking Time Seriously: A Theory of Socioemotional Selectivity," *American Psychologist* 54, no. 3 (1999): 165–81.

28. Angela M. Legg and Kate Sweeny, "Do You Want the Good News or the Bad News First? The Nature and Consequences of News Order Preferences," *Personality and Social Psychology Bulletin* 40, no. 3 (2014): 279–88; Linda L. Marshall and Robert F. Kidd, "Good News or Bad News First?" *Social Behavior and Personality* 9, no. 2 (1981): 223–26.

29. Angela M. Legg and Kate Sweeny, "Do You Want the Good News or the Bad News First? The Nature and Consequences of News Order Preferences," *Personality and Social Psychology Bulletin* 40, no. 3 (2014): 279–88.

30. See, e.g., William T. Ross, Jr., and Itamar Simonson, "Evaluations of Pairs of Experiences: A Preference for Happy Endings," *Journal of Behavioral Decision Making* 4, no. 4 (1991): 273–82. This preference is not uniformly positive. For example, people at the racetrack tend to bet more on longshots on the last race of the day. They hope to end with a bang but usually just end with emptier pockets. Craig R. M. McKenzie et al., "Are Longshots Only for Losers? A New Look at the Last Race Effect," *Journal of Behavioral Decision Making* 29, no. 1 (2016): 25–36. See also Martin D. Vestergaard and Wolfram Schultz, "Choice Mechanisms for Past, Temporally Extended Outcomes," *Proceedings of the Royal Society B* 282, no. 1810 (2015): 20141766.

31. Ed O'Brien and Phoebe C. Ellsworth, "Saving the Last for Best: A Positivity Bias for End Experiences," *Psychological Science* 23, no. 2 (2012): 163–65.

32. Robert McKee, *Story: Substance, Structure, Style, and the Principles of Screenwriting* (New York: ReaganBooks/HarperCollins, 1997), 311.

33. John August, "Endings for Beginners," *Scriptnotes* podcast 44, July 3, 2012, available at http://scriptnotes.net/endings-for-beginners.

34. Hal Hershfield et al., "Poignancy: Mixed Emotional Experience in the Face of Meaningful Endings," *Journal of Personality and Social Psychology* 94, no. 1 (2008): 158–67.

CHAPTER 5. TIME HACKER'S HANDBOOK

1. Jon Bischke, "Entelo Study Shows When Employees Are Likely to Leave Their Jobs," October 6, 2014, available at https://blog.entelo.com/new-entelo-study-shows-when-employees-are-likely-to-leave-their-jobs.

2. Robert I. Sutton, *Good Boss, Bad Boss: How to Be the Best . . . and Learn from the Worst* (New York: Business Plus/Hachette, 2010). That awful boss might also be miserable herself. See Trevor Foulk et al., "Heavy Is the Head That Wears the Crown: An Actor-Centric Approach to Daily Psychological Power, Abusive Leader Behavior, and Perceived Incivility," *Academy of Management Journal* 60, forthcoming.

NOTES

3. Patrick Gillespie, "The Best Time to Leave Your Job Is . . . ," *CNN Money*, May 12, 2016, available at http://money.cnn.com/2016/05/12/news/economy/best-time-to-leave-your-job/.

4. Peter Boxall, "Mutuality in the Management of Human Resources: Assessing the Quality of Alignment in Employment Relationships," *Human Resource Management Journal* 23, no. 1 (2013): 3–17; Mark Allen Morris, "A Meta-Analytic Investigation of Vocational Interest-Based Job Fit, and Its Relationship to Job Satisfaction, Performance, and Turnover," PhD diss., University of Houston, 2003; Christopher D. Nye et al., "Vocational Interests and Performance: A Quantitative Summary of over 60 Years of Research," *Perspectives on Psychological Science* 7, no. 4 (2012): 384–403.

5. Deborah Bach, "Is Divorce Seasonal? UW Research Shows Biannual Spike in Divorce Filings," *UW Today*, August 21, 2016, available at http://www.washington.edu/news/2016/08/21/is-divorce-seasonal-uw-research-shows-biannual-spike-in-divorce-filings/.

6. Claire Sudath, "This Lawyer Is Hollywood's Complete Divorce Solution," *Bloomberg Businessweek*, March 2, 2016.

7. Teresa Amabile and Steven Kramer, *The Progress Principle: Using Small Wins to Ignite Joy, Engagement, and Creativity at Work* (Boston: Harvard Business Review Press, 2011).

8. Jesse Singal, "How to Maximize Your Vacation Happiness," *New York*, July 5, 2015.

CHAPTER 6. SYNCHING FAST AND SLOW

1. Suketu Mehta, *Maximum City: Bombay Lost and Found* (New York: Vintage, 2009), 264.

2. Ian R. Bartky, *Selling the True Time: Nineteenth-Century Timekeeping in America* (Stanford, CA: Stanford University Press, 2000).

3. Deborah G. Ancona and Chee-Leong Chong, "Timing Is Everything: Entrainment and Performance in Organization Theory," *Academy of Management Proceedings* 1992, no. 1 (1992): 166–69. This line of thinking was presaged by Joseph McGrath, a University of Michigan social psychologist, in Joseph E. McGrath, "Continuity and Change: Time, Method, and the Study of Social Issues," *Journal of Social Issues* 42, no. 4 (1986): 5–19; Joseph E. McGrath and Janice R. Kelly, *Time and Human Interaction: Toward a Social Psychology of Time* (New York: Guilford Press, 1986); and Joseph E. McGrath and Nancy L. Rotchford, "Time and Behavior in Organizations," in L. L. Cummings and Barry M. Staw, eds., *Research in Organizational Behavior* 5 (Greenwich, CT: JAI Press, 1983), 57–101.

4. Ken-Ichi Honma, Christina von Goetz, and Jürgen Aschoff, "Effects of Restricted Daily Feeding on Freerunning Circadian Rhythms in Rats," *Physiology & Behavior* 30, no. 6 (1983): 905–13.

5. Ancona defines organizational entrainment as "the adjustment or moderation of one behavior either to synchronize or to be in rhythm with another behavior" and maintains that it can "be conscious, subconscious, or instinctive."

6. Till Roenneberg, *Internal Time: Chronotypes, Social Jet Lag, and Why You're So Tired* (Cambridge, MA: Harvard University Press, 2012), 249.

7. Ya-Ru Chen, Sally Blount, and Jeffrey Sanchez-Burks, "The Role of Status Differentials in Group Synchronization" in Sally Blount, Elizabeth A. Mannix, and Margaret Ann

Neale, eds., *Time in Groups*, vol. 6 (Bingley, UK: Emerald Group Publishing, 2004), 111–13.

8. Roy F. Baumeister and Mark R. Leary, "The Need to Belong: Desire for Interpersonal Attachments as a Fundamental Human Motivation," *Psychological Bulletin* 117, no. 3 (1995): 497–529.

9. See C. Nathan DeWall et al., "Belongingness as a Core Personality Trait: How Social Exclusion Influences Social Functioning and Personality Expression," *Journal of Personality* 79, no. 6 (2011): 1281–1314.

10. Dan Mønster et al., "Physiological Evidence of Interpersonal Dynamics in a Cooperative Production Task," *Physiology & Behavior* 156 (2016): 24–34.

11. Michael Bond and Joshua Howgego, "I Work Therefore I Am," *New Scientist* 230, no. 3079 (2016): 29–32.

12. Oday Kamal, "What Working in a Kitchen Taught Me About Teams and Networks," *The Ready*, April 1, 2016, available at https://medium.com/the-ready/schools -don-t-teach-you-organization-professional-kitchens-do-7c6cf5145c0a#.jane98bnh.

13. Michael W. Kraus, Cassy Huang, and Dacher Keltner, "Tactile Communication, Cooperation, and Performance: An Ethological Study of the NBA," *Emotion* 10, no. 5 (2010): 745–49.

14. Björn Vickhoff et al., "Music Structure Determines Heart Rate Variability of Singers," *Frontiers in Psychology* 4 (2013): 1–16.

15. James A. Blumenthal, Patrick J. Smith, and Benson M. Hoffman, "Is Exercise a Viable Treatment for Depression?" *ACSM's Health & Fitness Journal* 16, no. 4 (2012): 14–21.

16. Daniel Weinstein et al., "Singing and Social Bonding: Changes in Connectivity and Pain Threshold as a Function of Group Size," *Evolution and Human Behavior* 37, no. 2 (2016): 152–58; Bronwyn Tarr, Jacques Launay, and Robin I. M. Dunbar, "Music and Social Bonding: 'Self-Other' Merging and Neurohormonal Mechanisms," *Frontiers in Psychology* 5 (2014), 1–10; Björn Vickhoff et al., "Music Structure Determines Heart Rate Variability of Singers," *Frontiers in Psychology* 4 (2013): 1–16.

17. Stephen M. Clift and Grenville Hancox, "The Perceived Benefits of Singing: Findings from Preliminary Surveys of a University College Choral Society," *Perspectives in Public Health* 121, no. 4 (2001): 248–56; Leanne M. Wade, "A Comparison of the Effects of Vocal Exercises/Singing Versus Music-Assisted Relaxation on Peak Expiratory Flow Rates of Children with Asthma," *Music Therapy Perspectives* 20, no. 1 (2002): 31–37.

18. Daniel Weinstein et al., "Singing and Social Bonding: Changes in Connectivity and Pain Threshold as a Function of Group Size," *Evolution and Human Behavior* 37, no. 2 (2016): 152–58; Gene D. Cohen et al., "The Impact of Professionally Conducted Cultural Programs on the Physical Health, Mental Health, and Social Functioning of Older Adults," *Gerontologist* 46, no. 6 (2006): 726–34.

19. Christina Grape et al., "Choir Singing and Fibrinogen: VEGF, Cholecystokinin and Motilin in IBS Patients," *Medical Hypotheses* 72, no. 2 (2009): 223–25.

20. R. J. Beck et al., "Choral Singing, Performance Perception, and Immune System Changes in Salivary Immunoglobulin A and Cortisol," *Music Perception* 18, no. 1 (2000): 87–106.

21. Daisy Fancourt et al., "Singing Modulates Mood, Stress, Cortisol, Cytokine and Neuropeptide Activity in Cancer Patients and Carers," *Ecancermedicalscience* 10 (2016): 1–13.

22. Daniel Weinstein et al., "Singing and Social Bonding: Changes in Connectivity and Pain Threshold as a Function of Group Size," *Evolution and Human Behavior* 37, no. 2 (2016): 152–58; Daisy Fancourt et al., "Singing Modulates Mood, Stress, Cortisol, Cytokine and Neuropeptide Activity in Cancer Patients and Carers," *Ecancermedicalscience* 10 (2016): 1–13; Stephen Clift and Grenville Hancox, "The Significance of Choral Singing for Sustaining Psychological Wellbeing: Findings from a Survey of Choristers in England, Australia and Germany," *Music Performance Research* 3, no. 1 (2010): 79–96; Stephen Clift et al., "What Do Singers Say About the Effects of Choral Singing on Physical Health? Findings from a Survey of Choristers in Australia, England and Germany," paper presented at the 7th Triennial Conference of the European Society for the Cognitive Sciences of Music, Jyväskylä, Finland, 2009.

23. Ahmet Munip Sanal and Selahattin Gorsev, "Psychological and Physiological Effects of Singing in a Choir," *Psychology of Music* 42, no. 3 (2014): 420–29; Lillian Eyre, "Therapeutic Chorale for Persons with Chronic Mental Illness: A Descriptive Survey of Participant Experiences," *Journal of Music Therapy* 48, no. 2 (2011): 149–68; Audun Myskja and Pål G. Nord, "The Day the Music Died: A Pilot Study on Music and Depression in a Nursing Home," *Nordic Journal of Music Therapy* 17, no. 1 (2008): 30–40; Betty A. Baily and Jane W. Davidson, "Effects of Group Singing and Performance for Marginalized and Middle-Class Singers," *Psychology of Music* 33, no. 3 (2005): 269–303; Nicholas S. Gale et al., "A Pilot Investigation of Quality of Life and Lung Function Following Choral Singing in Cancer Survivors and Their Carers," *Ecancermedicalscience* 6, no. 1 (2012): 1–13.

24. Jane E. Southcott, "And as I Go, I Love to Sing: The Happy Wanderers, Music and Positive Aging," *International Journal of Community Music* 2, no. 2–3 (2005): 143–56; Laya Silber, "Bars Behind Bars: The Impact of a Women's Prison Choir on Social Harmony," *Music Education Research* 7, no. 2 (2005): 251–71.

25. Nick Alan Joseph Stewart and Adam Jonathan Lonsdale, "It's Better Together: The Psychological Benefits of Singing in a Choir," *Psychology of Music* 44, no. 6 (2016): 1240–54.

26. Bronwyn Tarr et al., "Synchrony and Exertion During Dance Independently Raise Pain Threshold and Encourage Social Bonding," *Biology Letters* 11, no. 10 (2015).

27. Emma E. A. Cohen et al., "Rowers' High: Behavioural Synchrony Is Correlated with Elevated Pain Thresholds," *Biology Letters* 6, no. 1 (2010): 106–108.

28. Daniel James Brown, *The Boys in the Boat: Nine Americans and Their Epic Quest for Gold at the 1936 Berlin Olympics* (New York: Penguin Books, 2014), 48.

29. Sally Blount and Gregory A. Janicik, "Getting and Staying In-Pace: The 'In-Synch' Preference and Its Implications for Work Groups," in Harris Sondak, Margaret Ann Neale, and E. Mannix, eds., *Toward Phenomenology of Groups and Group Membership*, vol. 4 (Bingley, UK: Emerald Group Publishing, 2002), 235–66; see also Reneeta Mogan, Ronald Fischer, and Joseph A. Bulbulia, "To Be in Synchrony or Not? A Meta-Analysis of Synchrony's Effects on Behavior, Perception, Cognition and Affect," *Journal of Experimental Social Psychology* 72 (2017): 13–20; Sophie Leroy et al., "Synchrony Preference: Why Some People Go with the Flow and Some Don't," *Personnel Psychology* 68, no. 4 (2015): 759–809.

30. Stefan H. Thomke and Mona Sinha, "The Dabbawala System: On-Time Delivery,

Every Time," Harvard Business School case study, 2012, available at http://www.hbs .edu/faculty/Pages/item.aspx?num=38410.

31. Bahar Tunçgenç and Emma Cohen, "Interpersonal Movement Synchrony Facilitates Pro-Social Behavior in Children's Peer-Play," *Developmental Science* (forthcoming).

32. Bahar Tunçgenç and Emma Cohen, "Movement Synchrony Forges Social Bonds Across Group Divides," *Frontiers in Psychology* 7 (2016): 782.

33. Tal-Chen Rabinowitch and Andrew N. Meltzoff, "Synchronized Movement Experience Enhances Peer Cooperation in Preschool Children," *Journal of Experimental Child Psychology* 160 (2017): 21–32.

CHAPTER 6. TIME HACKER'S HANDBOOK

1. Duncan Watts, "Using Digital Data to Shed Light on Team Satisfaction and Other Questions About Large Organizations," *Organizational Spectroscope*, April 1, 2016, available at https://medium.com/@duncanjwatts/the-organizational-spectroscope -7f9f239a897c.

2. Gregory M. Walton and Geoffrey L. Cohen, "A Brief Social-Belonging Intervention Improves Academic and Health Outcomes of Minority Students," *Science* 331, no. 6023 (2011): 1447–51; Gregory M. Walton et al., "Two Brief Interventions to Mitigate a 'Chilly Climate' Transform Women's Experience, Relationships, and Achievement in Engineering," *Journal of Educational Psychology* 107, no. 2 (2015): 468–85.

3. Lily B. Clausen, "Robb Willer: What Makes People Do Good?" *Insights by Stanford Business*, November 16, 2015, available at https://www.gsb.stanford.edu/insights /robb-willer-what-makes-people-do-good.

CHAPTER 7. THINKING IN TENSES

1. It's not 100 percent certain Groucho said this, either. See Fred R. Shapiro, *The Yale Book of Quotations* (New Haven, CT: Yale University Press, 2006), 498.

2. Anthony G. Oettinger, "The Uses of Computers in Science," *Scientific American* 215, no. 3 (1966): 161–66.

3. Frederick J. Crosson, *Human and Artificial Intelligence* (New York: Appleton-Century-Crofts, 1970), 15.

4. Fred R. Shapiro, *The Yale Book of Quotations* (New Haven, CT: Yale University Press, 2006), 498.

5. "The Popularity of 'Time' Unveiled," *BBC News*, June 22, 2006, available at http:// news.bbc.co.uk/2/hi/uk_new/5104778.stm. Alan Burdick also makes this point in his insightful book on time. See Alan Burdick, *Why Time Flies: A Mostly Scientific Investigation* (New York: Simon & Schuster, 2017), 25.

6. For a fascinating account of the history of nostalgia, and the sources of these quotations, see Constantine Sedikides et al., "To Nostalgize: Mixing Memory with Affect and Desire," *Advances in Experimental Social Psychology* 51 (2015): 189–273.

7. Tim Wildschut et al., "Nostalgia: Content, Triggers, Functions," *Journal of Personality and Social Psychology* 91, no. 5 (2006): 975–93.

8. Clay Routledge et al., "The Past Makes the Present Meaningful: Nostalgia as an Existential Resource," *Journal of Personality and Social Psychology* 101, no. 3 (2011): 638–22; Wijnand A. P. van Tilburg, Constantine Sedikides, and Tim Wildschut, "The Mnemonic Muse: Nostalgia Fosters Creativity Through Openness to Experience," *Journal of Experimental Social Psychology* 59 (2015): 1–7.

9. Wing-Yee Cheung et al., "Back to the Future: Nostalgia Increases Optimism," *Personality and Social Psychology Bulletin* 39, no. 11 (2013): 1484–96; Xinyue Zhou et al., "Nostalgia: The Gift That Keeps on Giving," *Journal of Consumer Research* 39, no. 1 (2012): 39–50; Wijnand A. P. van Tilburg, Eric R. Igou, and Constantine Sedikides, "In Search of Meaningfulness: Nostalgia as an Antidote to Boredom," *Emotion* 13, no. 3 (2013): 450–61.

10. Xinyue Zhou et al., "Heartwarming Memories: Nostalgia Maintains Physiological Comfort," *Emotion* 12, no. 4 (2012): 678–84; Rhiannon N. Turner et al., "Combating the Mental Health Stigma with Nostalgia," *European Journal of Social Psychology* 43, no. 5 (2013): 413–22.

11. Matthew Baldwin, Monica Biernat, and Mark J. Landau, "Remembering the Real Me: Nostalgia Offers a Window to the Intrinsic Self," *Journal of Personality and Social Psychology* 108, no. 1 (2015): 128–47.

12. Daniel T. Gilbert and Timothy D. Wilson, "Prospection: Experiencing the Future," *Science* 317, no. 5843 (2007): 1351–54.

13. M. Keith Chen, "The Effect of Language on Economic Behavior: Evidence from Savings Rates, Health Behaviors, and Retirement Assets," *American Economic Review* 103, no. 2 (2013): 690–731.

14. Ibid.

15. The conversation began with Edward Sapir, "The Status of Linguistics as a Science," *Language* 5, no. 4 (1929): 207–14. That view was discredited by, among others, Noam Chomsky, *Syntactic Structures*, 2nd. ed. (Berlin and New York: Mouton de Gruyter, 2002), only to be reconsidered again more recently. See, e.g., John J. Gumperz and Stephen C. Levinson, "Rethinking Linguistic Relativity," *Current Anthropology* 32, no. 5 (1991): 613–23; Martin Pütz and Marjolyn Verspoor, eds., *Explorations in Linguistic Relativity*, vol. 199 (Amsterdam and Philadelphia: John Benjamins Publishing, 2000).

16. See Hal E. Hershfield, "Future Self-Continuity: How Conceptions of the Future Self Transform Intertemporal Choice," *Annals of the New York Academy of Sciences* 1235, no. 1 (2011): 30–43.

17. Daphna Oyserman, "When Does the Future Begin? A Study in Maximizing Motivation," *Aeon*, April 22, 2016, available at https://aeon.co/ideas/when-does-the-future -begin-a-study-in-maximising-motivation. See also Neil A. Lewis, Jr., and Daphna Oyserman, "When Does the Future Begin? Time Metrics Matter, Connecting Present and Future Selves," *Psychological Science* 26, no. 6 (2015): 816–25; Daphna Oyserman, Deborah Bybee, and Kathy Terry, "Possible Selves and Academic Outcomes: How and When Possible Selves Impel Action," *Journal of Personality and Social Psychology* 91, no. 1 (2006): 188–204; Daphna Oyserman, Kathy Terry, and Deborah Bybee, "A Possible Selves Intervention to Enhance School Involvement," *Journal of Adolescence* 25, no. 3 (2002): 313–26.

18. Ting Zhang et al., "A 'Present' for the Future: The Unexpected Value of Rediscovery," *Psychological Science* 25, no. 10 (2014): 1851–60.

19. Dacher Keltner and Jonathan Haidt, "Approaching Awe, a Moral, Spiritual, and Aesthetic Emotion," *Cognition & Emotion* 17, no. 2 (2003): 297–314.

20. Melanie Rudd, Kathleen D. Vohs, and Jennifer Aaker, "Awe Expands People's Perception of Time, Alters Decision Making, and Enhances Well-Being," *Psychological Science* 23, no. 10 (2012): 1130–36. Helping others also expands our feelings of time, increasing our sense of "time affluence"; see Cassie Mogilner, Zoë Chance, and Michael I. Norton, "Giving Time Gives You Time," *Psychological Science* 23, no. 10 (2012): 1233–38.

INDEX

Daniel H. Pink has been thinking about you.

About how you can be more creative.

How you can motivate yourself and the people in your life.

How to sell your product, your idea, or yourself more effectively—and more ethically.

And how to calibrate the elements of timing in your life to benefit you at work and at home.

He's collected new information and cutting-edge scientific research and translated these insights into practical, human terms.

Because Pink understands that as the world changes, you must, too. So he writes books that give you the world—along with the tools and tips you need to work smarter and live better.

An exciting—and encouraging—exploration of creativity

The future belongs to a new kind of thinker with a different kind of mind, Daniel H. Pink argues in *A Whole New Mind*.

Drawing on fascinating research from around the world, both in the lab and in the workplace, Pink shines light on exciting ideas about creativity as he reveals that it's the creative and empathic right-brain thinkers who will create the future—one that's already here.

The *New York Times* and *BusinessWeek* Bestseller

"THIS BOOK IS A MIRACLE. Completely original and profound."
—Tom Peters, author of *In Search of Excellence*

UPDATED WITH NEW MATERIAL

A WHOLE NEW MIND

WHY RIGHT-BRAINERS WILL RULE THE FUTURE

DANIEL H. PINK

"Right on the money." —*U.S. News & World Report*

"A road map to help the rest of us guide our own pitches." —*Chicago Tribune*

An insightful career guide in *manga*

Meet Johnny Bunko. He's a young man who did what people told him to and got a job. But now he suspects that what he thought he knew about work is just plain wrong.

Daniel H. Pink reinvents the career guide as a smart and engaging story told through the Japanese comic art form of manga. Packed with smart, life-changing advice, this is a refreshingly different business book—and the last career guide you'll ever need.

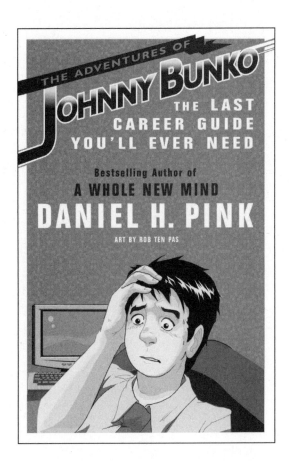

"Zen-like . . . witty." —*Time*

"Outrageous, delightful." —*The Wall Street Journal*

A bold approach to what motivates us

In *A Whole New Mind,* Daniel H. Pink wrote about the *what* of work, the traits essential for professional success and personal fulfillment. In *Drive*, Pink explores the *why* of work, our human need to direct our own lives, to learn and to create, and to better ourselves. Pink identifies the three elements of enduring motivation and then provides the tools to help you put them into place at work, at school, and at home.

NEW YORK TIMES BESTSELLER

"Provocative and fascinating." —MALCOLM GLADWELL

Daniel H. Pink

author of *A Whole New Mind*

DRIVE

The Surprising Truth
About What Motivates Us

"These are lessons worth repeating, and if more companies feel emboldened to follow Mr. Pink's advice, so much the better." —*The Wall Street Journal*

"Pink's ideas deserve a wide hearing." —*Forbes*

The power of selling: It's in all of us

Today 1 in 9 people are in sales. But dig deeper and a startling truth emerges: So are the other eight. That's right: No matter what we do for a living, we're all in sales now. And thanks to a cluster of economic forces, sales isn't what it used to be. In *To Sell Is Human*, Daniel H. Pink shows how everyone who expends effort moving others—convincing clients, colleagues, students and family members—is in sales. He explores the new science of selling, details the essential qualities we need to master to sell effectively, and lays out the key abilities for thriving in this new environment.

"Vastly entertaining and informative."—FORBES.COM

THE #1 *NEW YORK TIMES* BUSINESS BESTSELLER

AUTHOR OF *DRIVE* AND *A WHOLE NEW MIND*

DANIEL H. PINK

TO SELL IS HUMAN

THE SURPRISING TRUTH
ABOUT MOVING OTHERS

"Artfully blend(s) anecdotes, insights, and studies from the social sciences into a frothy blend of utility and entertainment." —***Bloomberg***

"A road map to help the rest of us guide our own pitches." —***Chicago Tribune***

A dynamic look at the transforming power of our most misunderstood emotion: regret

Everybody has regrets. They are part of life. Yet few of us know that we can use them to make better decisions and live happier lives. In *The Power of Regret*, Pink shows us how we can transform our regrets into a positive force for working smarter and living better.

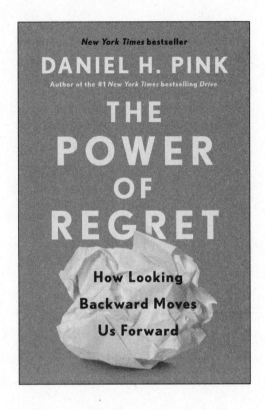

"A beautifully fragmented look at man's longing for permanence . . . Ambitious and complex." **—*The Wall Street Journal***

"The world needs this book." **—Brené Brown**